Motor Fleet Safety and Security Management

Second Edition

Motor Fleet Safety and Security Management

Second Edition

Daniel E. Della-Giustina

CRC Press is an imprint of the
Taylor & Francis Group, an **informa** business

CRC Press
Taylor & Francis Group
6000 Broken Sound Parkway NW, Suite 300
Boca Raton, FL 33487-2742

First issued in paperback 2017

No claim to original U.S. Government works
Version Date: 20120314

ISBN 13: 978-1-138-07249-7 (pbk)
ISBN 13: 978-1-4398-9507-8 (hbk)

Library of Congress Cataloging-in-Publication Data

Della-Giustina, Daniel.
Motor fleet safety and security management / Daniel E. Della-Giustina. -- 2nd ed.
p. cm.
Includes bibliographical references and index.
ISBN 978-1-4398-9507-8 (hardback)
1. Motor vehicle fleets--Safety measures. 2. Motor vehicle fleets--Security measures. 3. Traffic safety. I. Title.

HE5614.D375 2012
388.3'40684--dc23 2012008936

Visit the Taylor & Francis Web site at
http://www.taylorandfrancis.com

and the CRC Press Web site at
http://www.crcpress.com

Dedication

This book is dedicated to the beloved Della-Giustina grandchildren:

Dana, Daniel, Daniella, Denise, James, Katey, Marissa, Robyn, Steven, and Elisa.

Contents

Preface

The basic approach of the book is to cover the underlying concepts with the basic techniques and principles of motor fleet safety and security management.

In the wake of the September 11, 2001 attacks on the United States, businesses must now become proactive in their own security and planning for emergencies. The book contains approximately 40 percent more content than can be covered in one semester. With the new methods of transportation, travel has become much faster. Motor fleet use is advantageous because it is a quick method of delivering goods, which in turn strengthens the economy. Today, government, industry, and the private and public sectors, along with academia, must take a much more serious approach to the transportation of goods.

Loss of life is not the only loss associated with these incidents. The cost of injuries, lost wages, hiring and training expenses, insurance premiums, and property damage must be considered as well. With the risk of loss being so great, an effective transportation training program must be implemented in all motor fleet safety programs. Therefore, we can say that employers are primarily responsible for what happens to their drivers in the workplaces and terminals in the United States. It is the employers' responsibility to provide a safe place for their employees to work, along with safe work systems, safe vehicles, and adequate training and supervision. Employers must also ensure that whatever risk is created within the establishment will not spill over in the outside world to the detriment of the general public.

Motor fleet managers are concentrating their efforts toward accident prevention in all aspects of safety. Accident cost is a controllable business expense if handled correctly and monitored closely. Management is concentrating its efforts on controlling the direct and indirect costs of accidents and the injuries associated with them. Every motor fleet manager has a duty to provide safe transportation to his or her company's customers and employees, whether carrying passengers or transporting hazardous materials.

The goal of this book is to provide the reader with an understanding of a comprehensive motor fleet safety and instructional program. Topics covered in this book include the following:

- the elements of a fleet safety program
- accident prevention
- consideration for small-fleet driver selection, training, instruction, and supervision
- vehicle inspection
- how to organize accident data
- job safety analysis (JSA) safety meetings for commercial drivers
- fleet transportation publicity
- school bus safety

- shipping and storage of hazardous materials
- security in transportation

Chapter 15 contains a model of a motor fleet transportation program designed to assist in the implementation of a program. Each of these topics plays an important role in motor fleet safety.

To develop an understanding of the significance of mass transportation and motor fleets in today's society, the next time you are driving on a main highway, think about how many trucks are traveling on the road with you. I think you will agree that the amount of goods being transported every day is astronomical.

To transport goods and persons safely, everyone involved in the transportation system should be adequately trained, supervised, and monitored. Proper data and incident reports must be maintained to identify trends and problem areas. I hope this book provides you with the groundwork necessary to manage your motor fleet transportation system.

Acknowledgments

I am particularly thankful for the assistance and the resource information needed to support the contents of this text from M. A. Malik (graduate student), who assisted me in some editing of this text. I would like to express my gratitude to the five graduate students (B. Cunningham, C. Fotos, K. Ice, C. Pride, and T. C. Stewart) who were enrolled in our off-campus program in Wheeling, West Virginia, for providing a wealth of information to support this text. Also, I must give a profound thank you to Ron Swantek, safety manager, currently employed with FedEx Corporation in the Pittsburgh, Pennsylvania central office on the training aspect of motor fleet drivers.

For the development of the school bus safety chapter, I am indebted to Teresa Cole (CSP), safety director of the Central Region, First Student Corp., a National School Bus company, for sharing her professional insight and for affording me her always enthusiastic support. I also remain indebted to a large number of government agencies for the information they provided in both electronic and printed formats.

I would like to thank the many other people who contributed to this second edition, including the transportation practitioners and professional truck drivers who provided a wealth of information to support the contents of this text. There are so many others who have inspired me, and I want to acknowledge the reviewers of this book, several of whom are anonymous. Another key contributor, Robert Cutone, was extremely helpful with his information on the fire and explosive materials carried by the trucking industry on the U.S. highway system.

In addition, I would like to thank Dr. H. Ilkin Bilgesu, a tireless and patient fellow faculty member at WVU, who aided in the preparation of the text with his excellent computer skills.

Furthermore, I am forever indebted to Dr. Robert Nolan, former director of the Highway Traffic Safety Center at Michigan State University in East Lansing. Dr. Nolan was my doctoral committee chair and a loyal friend and mentor during my graduate studies.

About the Author

Daniel E. Della-Giustina holds a BA and an MA in education administration and psychology from the American International College at Springfield, Massachusetts. He earned an EdS in health and highway traffic safety and a PhD in higher education and educational research from Michigan State University. He has held two postdoctoral positions: one in environmental and emergency planning at Lancaster University in England and the other in wind engineering and environmental research at the University of Colorado. He is a certified hazard control manager.

Dr. Della-Giustina teaches in the Industrial and Management Systems Engineering Department, part of the Safety and Environmental Management Program, in the College of Engineering and Mineral Resources at West Virginia University. He has chaired and taught in the graduate program for 35 years. Concurrently, he has served as adjunct professor in the Petroleum and Natural Gas Engineering Department.

During Dr. Della-Giustina's tenure as chairman of the Safety and Environmental Management Department at West Virginia University, he built the program into the finest of its kind in the world. His seventeen textbooks in safety, health, and environmental disciplines are standard teaching references in U.S. and international educational institutions. In the corporate sector, his writings are used widely as the foundation upon which company-wide safety programs are built.

Dr. Della-Giustina developed safety management concepts that have wide applications. He alerted workers to the presence of hazardous substances through the use of material safety data sheets, worker training, product labeling, and other measures. He developed an instrument to investigate, analyze, and report accidents, including guidelines for processing workers' compensation claims. In concert with the Federal Emergency Management Agency (FEMA), he designed a tool for responding to emergency situations and the procedures for creating disaster recovery plans. He also created mechanisms to prevent fires and minimize losses involving hazardous materials.

While conducting a comprehensive national research project that included a critical evaluation of large and small school districts in all fifty states, Dr. Della-Giustina identified safety practices and needs in the transportation of handicapped students. His findings resulted in recommendations that were implemented by the U.S. Department of Transportation. Today, as a result of Dr. Della-Giustina's visionary work, all states have expanded their transportation programs for handicapped students (special populations) and adopted uniform safety standards.

In the early 1990s, Dr. Della-Giustina recognized the signs that violence in the workplace was going to increase. He began working to define the problem and seek methods of coping with what was to become a burgeoning national dilemma. Presently, his leadership is helping businesses understand potential warning signs of violence and determine ways to protect employees.

His accomplishments and contributions are legion. Dr. Della-Giustina was named a 2001 Hall of Fame award recipient at the National Safety Council's international convention in Atlanta, where he was recognized as an acclaimed leader and pioneer for his innovative contributions and service to the safety, health, and environment industries. Dr. Della-Giustina's most recent honor was in June 2005 when he was elected to the honor of Fellow in the American Society of Safety Engineers. These honors are the highest honors in each of these professional societies

Dr. Della-Giustina's family shares his commitment to safety. His wife, Janet, holds a master's degree in safety and environmental management and is a safety specialist in the Safety and Health Extension Services at West Virginia University. His son, Daniel, is a certified safety professional (CSP) and is the regional loss control manager for the Consigli Construction Corporation in Medford, Massachusetts. His other sons, David and John, both graduated from the U.S. Military Academy at West Point. David, a colonel, is director of emergency management at the Madigan Army Medical Center at Fort Lewis, Washington. John, a lieutenant colonel, is a counterintelligence officer at Fort Bliss, Texas. His stepson, Jacob Hyer, is a graduate of Wheaton College, Illinois.

1 Introduction

Technological advancements in the past have led to many changes to the world in which we live. Many of these changes have improved our society and quality of life. Some, however, have had detrimental effects on both humans and the environment. Transportation is no exception. It was not so long ago that people traveled by walking or by animal-powered means. Automobiles came along and made travel significantly faster. Giant tractor-trailers have the ability to move huge amounts of goods in a fraction of the time it once took. The transportation industry—specifically, motor fleets—operates more efficiently than ever before and continues to improve the means and safety of goods distribution.

These advancements are not without risks. Losses from transportation incidents are significant and include death, injury, wage losses, hiring and training expenses, higher insurance premiums, property damage, and business losses, just to name a few. By far, the greatest cause of accidental death in the United States involves motor vehicle incidents. About 50,000 people die and more than 2 million receive disabling injuries each year. The overall death rate is about 22 per 100,000 motor vehicle incidents. Although little attention is focused on the death rate from motor vehicle incidents while on the job, some studies suggest that between 25 and 35 percent of all job-related deaths involve motor vehicles.

LARGE TRUCKS

According to the National Safety Council's *Injury Facts* (2011), 3,380 fatalities from traffic crashes involving large trucks in 2009 represented a 20 percent decline from the 4,245 fatalities in 2008. More than 75 percent of these deaths were occupants of vehicles other than large trucks. Large trucks (10,000 lbs. and greater) are more likely to be involved in a multiple-vehicle fatal crash than passenger vehicles. In 2008, 82 percent of large trucks in fatal crashes were in multiple-vehicle crashes, compared to 58 percent of passenger vehicles. With two-vehicle crashes, 30 percent involved a large truck and another type of vehicle in 2008 with both vehicles being impacted in the front. The large truck was struck in the rear more than three times as often as the other vehicle (19 percent and 6 percent, respectively).

These statistics alone are proof of the incredible need to integrate safety into the management of the motor fleet industry. It not only makes good business sense to have a program in place to manage a motor fleet operation, but it is also a duty to provide safe transportation to the customer and the public as well.

Motor fleets, through their safety directors, are able to focus on both direct and indirect costs associated with their vehicular operations. Losses from motor fleet operations must be recognized as preventable incidents. A motor fleet safety program does for fleet operations what similar safety and loss control programs do for an organization: employ a competent, well-trained workforce; recognize hazards along with past losses; and take the appropriate action to prevent potential losses from occurring.

MOTOR FLEET INDUSTRY

The use of motor fleets and bus transportation has increased rapidly during the past four decades. Trucks transport most of the tonnage moved from one section of the country to another. The trucking industry has enabled all aspects of agriculture to expand that industry's spheres of contact. Recent safety records of buses and trucks have been markedly better than those of previous years based on a number of reasons, such as the commercial driver license (CDL) and effective training and educational safety programs for drivers.

Since World War II, both the commercial vehicle death rate and the accident rate per 100,000 vehicle miles have dropped greatly, with driver training being a key reason. How has this great safety record been improved?

To answer this question, we need to look at the Interstate Commerce Commission (ICC) and the Federal Motor Carrier Safety Administration (FMCSA), which have ensured the adoption of uniform safe practices and policies by the motor fleet and bus transit industries. Enterprises today train all new drivers and develop accident reports, with follow-up corrective measures and vehicle inspections as part of a daily program. Safety equipment and devices such as mechanical and electrical signals, side- and rear-marked lights and reflectors, along with hydraulic and air brakes, have been implemented. Well-organized programs in safe driving practices are required, and the hours of driving are defined and limited. Various enterprises hold meetings with drivers to discuss common problems and provide training. Competition among drivers and fleets has often resulted in an efficient motivational program to keep individual records clean. Incentives for best driving records sometimes take the form of promotions, bonuses, and other prizes. The fewer accidents and claims a motor fleet has, the lower its insurance premiums are.

The motor fleet industry has pioneered the development and implementation of traffic safety education programs. These programs have resulted in fleet and bus driver education, driver supervision, efficient terminal safety management, and have ultimately improved highway safety.

The motor carrier industry places a premium on safety. The motor fleet industry recognizes that achieving safe operations is an ongoing process that demands a number of different actions undertaken at all levels of an organization and fully integrated into hiring, compensation, and benefit decisions in day-to-day operating procedures. Almost every motor fleet operation in the United States has a safety management program in place. Safety managers are continually seeking ways to develop these programs and improve their safety performance.

A state is eligible to participate as a registration state and to receive revenue only if, as of January 1, 1991, it charged or collected a fee for a vehicle identification stamp or a number in line with the predecessor in this part of the standard. Each motor carrier is required to register and pay filing fees and must select a single participating state as its registration state. The carrier must select the state in which it maintains its principal place of business, if such state is a participating state. A carrier that maintains its principal place of business outside of a participating state must select the state in which it will operate the largest number of motor vehicles during the next registration year.

TRANSPORTATION OF HAZARDOUS MATERIALS

The transportation of hazardous materials and dangerous goods is widespread, with shipments of over 600,000 lbs. per week of these materials worldwide. Organizations involved with the transportation industry each day include the United Nations, the International Maritime Organization, the International Civil Aviation Organization, the International Air Transportation Association, the International Atomic Energy Agency, and the U.S. Department of Transportation. These organizations regulate the transportation of dangerous goods and hazardous materials. Accountability for properly identifying, understanding, and confirming the classification is among the shipper's responsibilities.

Under Title 49 of the Code of Federal Regulations (CFR) Parts 100 to 185 (49 CFR 100–185) are rules and regulations concerning the transport of hazardous materials. In Chapter 14, we discuss training, compliance, labeling, marking, packaging, classifications documentation, and the importance of certifying a shipment for transportation of hazardous materials via the U.S. highway system.

Rules and regulations established by the U.S. Department of Transportation (DOT) regarding the transportation of these materials are administered by the DOT Research and Special Programs Administration. All motor carrier safety programs must comply with the DOT regulations. The DOT sets and enforces these motor carrier transportation rules because it focuses on transportation activities and highway safety. (For more information, see Chapter 14, Shipping and Storage of Hazardous Materials.)

TRANSPORTATION SECURITY

In a changing world, work issues and transportation security concerns surround the growing number of vehicle shipments. These issues can affect the types and nature of risks present in the workplace, highways, and trucking terminals. All types of goods, especially motor vehicles transported to such points as the Newport News

Marine Terminal, San Francisco, and other locations, will be loaded onto car carriers and later shipped throughout the United States. In 2003, the Newport News Marine Terminal was expected to transport roughly 43,000 cars to various places between New York and Florida. These thousands of cars, each potentially carrying a bomb, are increasingly observed by national security specialists as a key focus in the battle against terrorism.

In efforts to prevent dirty bombs from coming to the United States by ship, national security agency leaders and port officials must focus primarily on inspecting motor vehicles before they are loaded onto car carriers. A bomb can potentially be hidden inside a trunk and activated either inside or outside the shipping terminal.

The Federal Transportation Security Administration acknowledges that, while its focus at the outset has indeed been on packages and containers, it now sees non-containerized cargo as a security concern. Stopping bombs from entering the United States is also a new concern for Distribution Auto Services, which runs many car importing centers. During the past several years, some workers have attempted to steal cars while driving them off the ships, intending to crash them through terminal fences. However, preventing bombs has been the main goal of this group since September 11, 2001. Its terminals will soon contain scanners mounted at heights of fifteen feet to detect radioactive explosives (fifteen feet above the ground surface is considered the optimal height for scanner effectiveness).

All motor fleet systems will be affected, and quality transportation security systems must be installed in the fight against terrorism. To guarantee these systems, we must understand and be familiar with all elements in the transportation security networks. We must analyze and respond to incoming threats.

To ensure viable transportation security programs, we must be prepared, knowledgeable, and familiar with security tools. Employing a wide variety of network devices and services is one of the key defense tactics in the security toolkit. In addition, establishing a security policy is important to addressing the risk. In the industrial arena, plants are likely to base security service judgments directly on vulnerability factors. For example, plants whose products are very small or valuable (such as jewelry, watches, and calculators) have a high risk factor for theft. In contrast, a home furnace/air-conditioning manufacturing plant might not be as concerned about theft because their products are so large. (See Chapters 15 and 16, which focus on security information.)

KEY ELEMENTS OF THIS TEXT

This particular book deals with the development of motor fleet transportation safety and identifies problems that are inherent to the industry. Some problems that are discussed are accident prevention, driver selection, driver training instruction, special considerations for a small fleet, school transportation, shipping and storage of hazardous materials, motor fleet accident data, and motor vehicle inspections. Other key elements of the text suggest methods that readers can use to combat these common problems. Although many things can be done to reduce these transportation management and security problems, the primary goal of this text is to increase readers' understanding of transportation safety and to suggest useful sources of information that will lay the foundation for their own motor fleet programs.

Management personnel consider the human aspect of loss and suffering, and they understand motor fleet safety and security in terms of dollars and cents. Investigation into the basis of losses may reveal sources that can be translated into monetary value. This is discussed in more detail in the next chapter.

REFERENCES

Baker, J.S., *Traffic Accident Investigation Manual*, Northwestern University Traffic Institute, Evanston, Illinois, 1986.

National Safety Council, *Injury Facts*, Itasca, Illinois, 2011.

2 Elements of a Fleet Safety Program

PURPOSE OF A SAFETY PROGRAM

A fleet safety program is designed to systematically reduce the number of on- and off-the-job work-related injuries. It should address management standards and policies; the recording of incidents and injuries; program results; employee selection, hiring, and training; and reward programs.

MAIN ELEMENTS OF A FLEET SAFETY PROGRAM

A fleet safety program consists of eleven main elements:[1]

1. Management leadership
2. A written safety policy
3. Safety responsibility assignments
4. Accident reports
5. Driver selection
6. Vehicle safety
7. Employee safety
8. Off-the-job safety
9. Safety supervision
10. Interest-sustaining activities
11. Integration of safety with the job

We look at each element in further detail.

Management Leadership

The success of any safety program depends on the support of top management. Communication, coordination, and cooperation among the organizations responsible for the roadway, human, and vehicle safety elements are a must. Management must establish short- and long-term highway safety goals to identify and address safety problems. Collecting, analyzing, and sharing highway safety data are key to ensuring that the best possible decisions are made. The active support of top management translates into employee commitment at all levels of the organization

When implementing a fleet safety program, management must take certain steps to ensure clarity and purpose.[2] The first step is to organize company accident reports to identify problems. This should include direct costs as well as workers' compensation, insurance, and medical costs. Once this information is organized, it should be used to develop a clear and precise company safety policy. After the policy is developed, the safety program is then set up and administered by a designated safety professional. The safety program should be that employee's sole responsibility. It is also important that all levels of management understand their roles in the safety program. It is every supervisor's responsibility to work with the safety professional to prevent vehicle accidents and employee injuries.

An analytical procedure that takes fault-tree analysis one step further is known as management oversight. It is used to analyze accidents to determine basis and contributory causes and a tool to evaluate existing safety programs important to the trucking industry. Ultimately, a postaccident investigation will usually indicate that there had been less than adequate management support in training and supervision. However, each safety manager must budget funds for all safety equipment necessary to every department of their organization. Funds should also be available for the training of supervisors and hourly employees, and for holding safety meetings on a regular basis. It is also important to have resources available for accident investigations, inspections, and audits. The most important budget consideration is ensuring that the accurate costs of all incidents are charged to the correct department or division.

Written Safety Policy

A company's outlook on safety should be communicated to employees, supervisors, and upper management through a written safety policy. This policy is useful in enforcing safety throughout the company. Formats of written policies will vary, but they should be clear and to the point. Sometimes a company's safety policy will include a motto or slogan. This motto is usually printed on various objects and displayed throughout the organization; every employee should be familiar with it.

According to the National Safety Council's *Motor Fleet Safety Manual*[8], management's attitude toward safety should be carefully phrased in a written safety policy communicated to all employees, supervisors, and top management. Also, a written policy can serve as a reference to settle issues of safety rules. The precise form of the written policy varies, but it should be a clear, forceful statement of management's desires.

Safety Responsibility Assignments

As mentioned earlier, upper management is ultimately responsible for the safety performance of its motor fleet. It is the responsibility of upper management to designate safety authority to managers and supervisors who communicate the responsibility down the chain to the drivers. This authority should be delegated to a full- or part-time safety director.

The designated safety director is responsible for directing accident prevention activities throughout line management. Some of his or her duties include organizing and administering an effective safety program; identifying existing hazards and developing plans to eliminate them; keeping records of fleet safety performance; and keeping management informed of program success. The ideal fleet safety director is knowledgeable of the industry, committed to safety, and a good administrator[9]. In a large fleet, this should be a full-time position. If a company has a small motor fleet, this position could be a part-time responsibility.

Managers and supervisors must be held accountable for the safety performance of their departments. To enforce accountability, safety performance should be a performance goal or objective that is a factor in salary raises or job promotions of all managers and supervisors.

Accident Reports

An accident prevention program should have a logical approach to gathering and applying accident information. This information is vital to managing claims. The following information is necessary to maintain complete accident investigation records.[1]

A well-developed accident prevention report must include a sound approach to gathering information on the fleet's accidents and implementing a plan to curtail future accidents. This is an important procedure to enhance the motor fleet operation in handling claims, especially for the insurance carriers. At a minimum, every report should include:

- Date of the accident
- Driver's name
- Driver's work location
- Vehicle number
- Accident type
- Accident location
- Objects, persons, or vehicles involved
- Estimated cost information
- Reportable or nonreportable
- Reports furnished

Driver Selection

When companies select drivers for their motor fleets, it is important that they perform thorough background checks of applicants' driving records.[3] This gives the company an idea of the risks or potential for loss if they hire these drivers. Selecting good drivers is the key to a program's success. Often, new drivers are experienced but they may have developed poor and potentially unsafe driving habits.

The establishment of a commercial driver's license (CDL) was passed by Congress on October 26, 1986, under the Commercial Motor Vehicle Safety Act of 1986. The act covers all the aspects of safety when transporting passengers and transporting commodities throughout the United States. The driving skill of the driver can be determined from an examination of his or her driving experience and a review of the CDL endorsements.

Vehicle Safety

Every company vehicle should be kept in safe condition and equipped with the proper registration, paperwork, and emergency equipment.[3] There should be customary pretrip and posttrip inspections, with clear procedures for reporting unsafe conditions. Procedures should also be established for all conceivable emergencies.

Employee Safety

A fleet safety program must have clear policies and procedures to prevent accidents and ensure employee safety. All employees should have their own roles specified in the safety program. Systematic accident reporting should be in place to prevent recurrence. All employees should be trained to follow safe working procedures and practices.

The organization must be designed to prevent accidents from occurring on its property. Buildings should have adequate lighting, climate control, and a good ventilation system. Housekeeping practices must be superior. Not only does this oversight help reduce injuries and illnesses, but it also creates an organizational culture that tells employees that management cares about their safety.

Off-the-Job Safety

It is management's responsibility to stress the importance of working safely off the job. Short safety talks can address the issue of safety precautions while working at home—for example, when mowing the lawn or using a ladder. Special rewards or incentives could be offered for safe behaviors off the job.

Safety Supervision

Employees' immediate supervisors or managers are the most important people in a safety program. They are the primary people who can motivate employees and set good examples for them to follow. Good safety managers must have support from top management to be successful.[4]

It is also the responsibility of supervisors or managers to identify and correct at-risk behaviors and practices. Safety standards must be met, and unsafe acts or conditions must be identified and corrected.[5] Management's failure to deal with these problems can eventually result in company losses.[6]

Interest-Sustaining Activities

A successful safety program keeps employees' attention. Rewards and recognition are the keys to doing this. There are several types of reward and recognition activities. Safety talks or meetings, posters, and bulletins provide information and continuously remind employees of the importance of working safely. Recognizing accident-free performance can give employees a goal for which to aim. Banquets, special events, and meetings involving management show an interest in group and individual safety efforts.

Integration of Safety with the Job

Safety should be integrated into every job—not just management jobs.[4] It should be stressed that the most effective way to do a job is the safe way. This ought to be stated clearly during the training process and carried out through an employee's entire career with the organization.

DUTIES OF FLEET SAFETY DIRECTORS

Investigation

Fleet safety directors are directly involved in all accident investigations. On a continual basis, they are responsible for examining company accident reports, existing work practices, facilities, and equipment to identify problem areas.[1] It is important for fleet safety directors to analyze past accident information to determine frequently occurring incidents and key problem areas. They should also visit all assigned facilities on a regular basis. Not only does this show interest and concern, it is also beneficial in observing work procedures, practices, and supervisor involvement.[6]

Planning

Planning involves the interpretation of accident data to develop a plan of action. This plan of action is designed to improve fleet safety performance. It may call for a change in current work practices and procedures, as well as training and supervision. It is important that this plan of action be flexible and open for management's comments and suggestions.[1] Revisions should be permitted and even expected.

Implementing the Program

Once the plan of action has been finalized, the fleet safety director must sell it to management to get its commitment to change.[1] These changes should be presented in a professional manner. Costs and benefits of the change should be identified. The fleet safety director should emphasize the impact the changes will have on direct and indirect accident costs, insurance costs, and employee production.[2]

Safety Training

Fleet safety directors are responsible for reviewing job training, required qualifications, and work procedures. The responsibility for conducting the training belongs to the safety supervisor or manager.[1] Fleet safety directors approve and support the training process. In some cases, they may be involved in the actual training.

Failure to provide adequate safety training for the driver might end up in the courts. The court might have to decide whether the employee knew before the accident that he/she was required to carry out an important safety task. By supporting the fact that the company's actions or inactions resulting in the lack of knowledge by the driver was the cause of the accident. The company will have to determine if the driver was negligent and, if so, whether negligence caused the accident. According to a number of court cases if the employer could be found negligent because it failed to prescribe, promulgate, and/or enforced adequate rules, procedures, policies, and regulations for the safe operation of the truck during the training of their drivers.

To prevent negligent claims by injured individuals, the company should make sure that all employees are adequately trained. You should introduce the most recent information and use safety training videos for new hires and also advanced training for all drivers. All of your training courses should be routinely reviewed and updated to reflect changes in the federal regulations or best practices in the trucking industry.

Program Follow-Up

Continual follow-up should include the following:[5]

- Monthly reports or memos of absenteeism
- Compliance issues
- Safety inspection results
- Incident report data

ELEMENTS OF A FLEET ACCIDENT PREVENTION PROGRAM

Why Gather Accident Information?

Gathering accident information is essential in identifying and eliminating root causes of accidents. It is also pertinent to a company for the following reasons: to obtain the best defense in court, to place the responsibility where it belongs, and to promote good labor relations.[1]

Valid and Reliable Sources of Reports

Since 1975 the National Highway Traffic Safety Administration (NHTSA) has been using a computerized database to analyze fatal motor vehicle crashes. This program, known as the Fatal Accident Reporting System (FARS), enhanced police accident reports, driver licensing files, and reports from medical examiners to determine the elements that contributed to the fatalities. This provided the statistical data on a wide range of safety-related problems to assist the trucking industry with information about accident trends.

Other sources of information to provide national data on occupational motor-vehicle-related crashes, injuries, and fatalities include:

- *Large Truck Crash Facts*—Federal Motor Carrier Safety Administration (FMCSA)
- *Survey of Occupational Injuries and Illnesses*—Bureau of Labor Statistics (BLS)
- *Census of Federal Occupational Injuries*—Bureau of Labor Statistics (BLS)

All of the above update their data on a regular basis.

The Accident Reporting Form

This form is the basic tool in an accident investigation. The accident reporting form identifies vital information necessary to analyze the accident.[3] There are several types of accident reporting forms. The best type is one that provides accurate information and is easy to complete. Any form can be modified to fit the needs of any organization. Once the report is complete, the person reading it should be able to visualize exactly what happened.[1]

Application of Accident Information

Accident information can be used to prevent future accidents of a similar nature. Accident reports should be prepared for management on a monthly basis.[7] Company incident rates, frequent accident locations, accident cost per mile, and hours of continual driving per accident should be included in this report.[1]

Accident data may also be useful in the selection of drivers. It is important to conduct background driving record checks on all potential employees. Accident data can also be useful in the training process.[2] It can uncover areas in which employees may need refresher training. It may also help identify areas in which drivers were never trained.

Safety talks or meetings are also good ways of presenting accident data. In this manner, accident prevention efforts can be targeted toward current and frequently occurring problem areas. In addition to all of these reasons, accident data is extremely important in rewarding and recognizing safe performance.

STUDY QUESTIONS

1. What is the purpose of a fleet safety program?
2. What are the eleven main elements of a fleet safety program?
3. List five pieces of necessary information in accident reporting.
4. (True or False) It is not necessary to do background checks of employees' driving records.
5. (True or False) The fleet safety director may be directly involved in training.
6. Why is training a key point when the term of negligence is determined during accidents?
7. Discuss the importance of having a written safety policy.
8. (True or False) Collecting, analyzing, and sharing highway safety data are key to ensuring that the best possible decisions are made.
9. Discuss why safety should be integrated into every job - not just management positions.
10. (True or False) Accident data and information can be used to prevent future accidents of a similar nature from occurring.

REFERENCES

1. Brodbeck, J.E., ed., *Motor Fleet Safety Manual*, 4th ed., National Safety Council, Itasca, Illinois, 7–50, 1996.
2. Moser, P., Rewards of creating a fleet safety culture, *Professional Safety*, American Society of Safety Engineers, Des Plaines, Illinois, 39–41, 2001.
3. http://www.FMCSA.dot.gov (accessed December 31, 2011).
4. http://www.safety.fhwa.dot.gov/fourthlevel/pro_res_safetymgt_key.htm (accessed December 31, 2011).
5. Baker, J.S., *Traffic Accident Investigation Manual*, Northwestern University Traffic Institute, Evanston, Illinois, 1986.
6. http://www.safety.fhwa.dot.gov/fourthlevel/pro_res_safetymgt_over.htm
7. *Motor Fleet Safety Supervision: Principles and Practices*, 4th ed., National Committee for Motor Fleet Supervisor Training, Alexandria, Virginia, 1987.
8. *Motor Fleet Safety Manual*, 4th ed., National Safety Council, Itasca, Illinois, 1996.
9. American Association of State Highway and Transportation Officials. *Manual for Assessing Safety Hardware* (MASH). Washington, D.C., 2009.

3 Accident Prevention

The National Highway Traffic Safety Administration (NHTSA) statistics for 2006 show that 106,000 people were injured in crashes with large trucks (10,001 pounds or greater). This was based on drivers who traveled 2.996 billion miles. However, in 2006 the overall fatality rate was down to a historic low of 1.45 per 100 million miles traveled, with 42,462 people killed and 2,575,000 people injured in motor vehicle accidents.[16] Vehicle accidents involving large trucks are of serious concern for motor fleet safety managers, with the data showing that more than 40 percent of all occupational fatalities are related to transportation. Moreover, accidents involving large trucks accounted for 11 percent of all traffic fatalities, or 4,995 deaths. Highway crashes were the leading cause of occupational fatalities in the United States (average of 1,300 civilian deaths each year, or 22 percent of all worker deaths).

REASONS FOR ACCIDENT PREVENTION

Moral Obligation

Employers have a moral obligation to prevent accidents. Employees are the lifelines of any organization. Their safety and health play crucial roles in an organization's ability to create products and services for its customers. When employees' safety and health are in jeopardy or are compromised through a lack of proper accident prevention, the organization itself becomes vulnerable.

Competition among motor fleet organizations can be fierce. An effectively managed accident prevention program can give an organization a competitive edge and help it to maintain or even increase its market share. Reducing accidents can substantially affect the bottom line or profit margin. The cost savings from lower insurance premiums and workers' compensation rates, along with increased productivity, can be reinvested into the organization.

An organization's safety culture can also help to attract better-qualified employees. Competition for the best qualified candidates extends past salary requirements and into benefits and organizational culture. Job seekers in today's market have begun to focus more attention on nonmonetary attributes of potential jobs. For instance, benefits packages have become valuable compensation tools to attract and retain employees. An organization's safety culture can accomplish the same result. A strong focus and commitment to safety demonstrates that a company values its employees.

Legal Obligation

Employers, according to the Occupational Safety and Health Act of 1970, Section 5(a)(1), "shall furnish to each of [their] employees employment and a place of employment which are free from recognized hazards that cause, or are likely to cause, death or serious physical harm."[2] More commonly referred to as the General Duty Clause, this statement provides a legally enforceable standard for protecting workers. If found in violation of the General Duty Clause, employers may be cited and receive substantial monetary penalties.

Financial Obligation

Financial obligations are another reason organizations need to implement effective accident prevention programs. An organization has a financial responsibility to its stockholders to do whatever is within its means to make the business profitable. An effective safety program—one that focuses on proactive accident prevention—can save the company money. The profit is recognized as savings from various expenditures, such as lower insurance premiums, lower workers' compensation rates, a decrease in equipment repairs, and other indirect costs.

Another financial obligation is to protect the organization from litigation resulting from accident injuries and fatalities. Liability insurance is a necessity in almost every form of business and is especially important for those companies with motor fleets. Accident victims and their families, as well as injured employees and their families, can file lawsuits. Therefore, accidents involving any injuries or fatalities threaten the viability of the organization.

MAIN AREAS OF FLEET ACCIDENTS

Vehicle Accidents

Motor vehicles have become the primary mode of transportation for people and products in the United States. Each year millions of drivers take to the roadways for business and leisure. In 2009, the National Highway Traffic Safety Administration (NHTSA) reported that drivers traveled 2.979 billion vehicle miles.[3] It is not surprising, therefore, that according to the NHTSA Fatality Analysis Reporting System (FARS) there were an estimated 6,394,000 police-reported traffic crashes in 2000.[4]

Motor vehicle collisions are the number one cause of preventable deaths and injuries—which is not unexpected given the millions of miles traveled each year. In 2009, although the overall fatality rate was down to a historic low of 1.21 per 100 million miles traveled, 35,900 people were killed and 3,500,000 received injuries requiring some form of medical consultation as a result of the motor vehicle accident.[4] Despite popular perceptions, the majority of accidents occur in industries that rely on a mobile sales force or local servicing or deliveries rather than long-haul trucking. However, motor fleet organizations can expect approximately 20 percent of their drivers to be involved in an annual vehicle accident. One in three will involve a fatality.

Vehicle accidents involving large trucks are of serious concern for motor fleet safety managers. Large trucks are more likely to be involved in a multiple-vehicle fatal crash than are passenger vehicles.[4] Moreover, accidents involving large trucks accounted for 11 percent of all traffic fatalities, or 4,719 deaths. Medium trucks (those with a gross vehicle weight rating of under 10,000 pounds) accounted for 562 deaths in 2000, according to NHTSA data. Large and medium truck accidents resulted in over 140,000 injuries. During 2001 a total of 5,082 people were killed and 131,000 were injured in crashes involving large trucks.

A 2002 research study prepared for the AAA Foundation for Traffic Safety found that automobile drivers contribute more to crashes involving large trucks than do commercial vehicle drivers. Some 75 percent of all driver errors were caused by automobile drivers, compared to 25 percent for truck drivers.[4]

Employee Injury Accidents

It is important to maintain employee injury or illness experiences of all employees by keeping information about all injuries (e.g., recordable illnesses, first-aid cases, and disabling injuries) in a log or computer for easy retrieval of the data. The reason for this type of record keeping is based on the premise that supervisors cannot be expected to remember the experiences of individual workers over time. Truck driver corrective action, such as retraining or reassigning to different tasks within the enterprise, may show consistent concern for the employee's own welfare or the safety of other workers.

In 2000, over 5.7 million employee injuries and illnesses were reported, with 2.8 million requiring time away from work beyond the day of the incident, restricted duties at work, or both, according to the Bureau of Labor Statistics (BLS).[5] Also, injuries and illnesses requiring only restricted work remained steady from previous years at approximately 1 million. Figure 3.1, based on data from the BLS, shows that of the ten occupations accounting for approximately one third of all injuries and illnesses requiring time away from work from 1993 to 1999, truck drivers reported the highest numbers since 1993.

The BLS data revealed that injuries and illnesses resulting in time away from work or restricted activity were caused by various reasons. Specifically, over 40 percent of all cases reported between 1993 and 1999 involved sprains, strains, or tears, most often of the back.[5] According to the BLS, overexertion while maneuvering

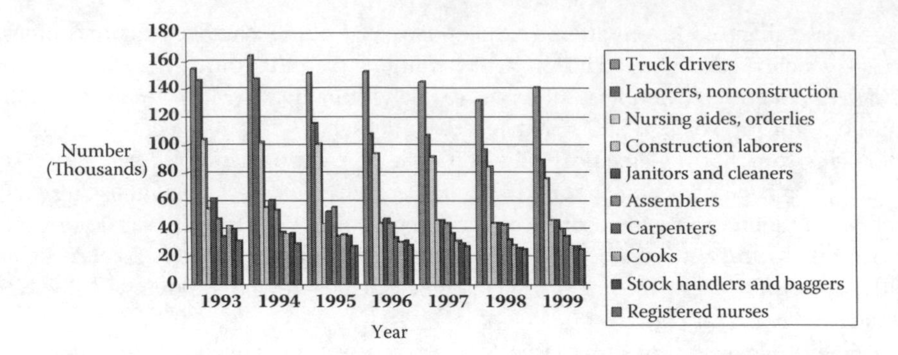

FIGURE 3.1 Occupational injuries and illnesses with time away from work. (From the Bureau of Labor Statistics, 1999). A motor fleet safety program should be designed to systematically reduce the number of on- and off-the-job work-related injuries. Accident cost is a controllable business expense. Management should concentrate its efforts toward controlling the direct and indirect costs of accidents and the injuries that are associated with them. Based on this factor, an accident prevention program is a key procedure to encompass a systematic approach to preventing accidents when the causes and risk factors are determined. Every motor fleet manager has a duty to provide safe transportation to its customers and employees.

objects and contact with objects and equipment were contributing factors leading to disabling events or exposure in 16 to 40 percent of cases reported overall. Other common reported causes for lost work time included bruises and contusions, cuts and lacerations, and fractures.

Off-the-Job Accidents

Most organizations do not consider the impact of off-the-job accidents because there are no reporting requirements associated with them. Statistics show that employees are safer on the job than they are on the highway driving their trucks or even in the home. Injuries to workers while off the job represent an additional expense to the motor fleet operation and are usually not included when developing an annual budget. However, this type of incident has a substantial effect on the safety of employees and is a source of significant costs. John C. Myre states that 60 percent of accidents that keep employees off the job occur away from work.[6] The National Safety Council reported that in 1995 over 88,000 people died from off-the-job injuries, with almost one half as a result of motor vehicle accidents. According to the National Safety Council's *Injury Facts*, nine out of ten deaths and more than two thirds of the disabling injuries suffered by workers in 2007 occurred off the job.[17] It is estimated that the lifetime odds of being killed in an accident off the job is about one in fifty.[6]

The chances of a person being temporarily or permanently disabled from an off-the-job accident are far greater than being killed from one. It is estimated that each year over 15 million people suffer some form of temporary or permanent disabling injuries while away from the job. In some instances, workers injured off the job report to work later and sustain additional injuries from performing work duties.

Such cases can have a dramatic effect on workers' compensation rates and other costs associated with occupational injuries and illnesses.

EXPENSE OF INJURIES AND ILLNESSES

Injuries and illnesses resulting from accidents are extremely costly, and many employers are unaware of what their human and financial costs actually are. Accidents are more expensive than most employers realize. Both direct and indirect components must be considered in the total costs for injuries and illnesses, which is difficult to compute, given that countless hidden costs are involved. In 1993 the economic cost for injuries alone from accidents was estimated at more than $110 billion.[4] According to the *Journal of Environmental Health*, work injury costs in 1994 rose to $121 billion in medical care, lost productivity, and wages.[7]

THE PYRAMID

The total costs of accidents are often depicted as a pyramid, as shown in Figure 3.2. The pyramid depicts the direct costs to reveal obvious costs, as well as workers' compensation claims, medical costs, and indemnity payments. Indirect costs, however, are less obvious, if realized at all. Such costs include schedule delays, added administrative time, lower morale, increased absenteeism, and poorer customer relations. Studies have shown that the ratio of indirect to direct costs varies widely from as high as 20:1 to as low as 1:1.[8] It is apparent from the data that indirect costs constitute the bulk of the total cost.

COST OF OFF-THE-JOB INJURIES

Off-the-job accident costs often exceed on-the-job accident costs in many organizations. In addition, some workers' compensation claims may actually begin as minor off-the-job injuries. According to Myre, each year American companies pay over $400 per employee to cover expenses such as health care costs resulting from these types of injuries.[6] Based on this cost estimate, an employer with 2,000 employees and a 10 percent profit margin needs $8 million in revenue just to pay for off-the-job

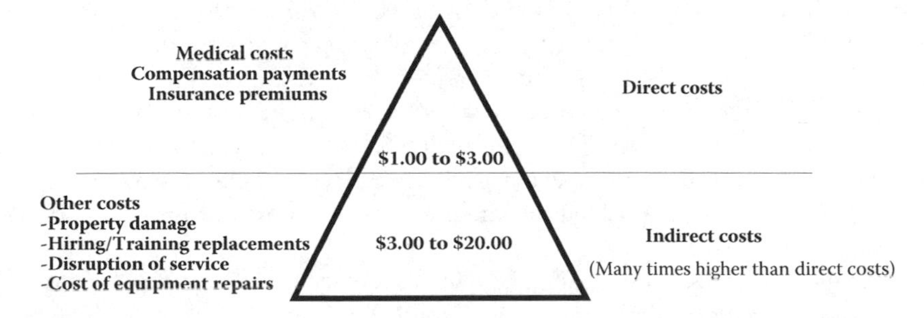

FIGURE 3.2 Cost of occupational injuries and illnesses with time away from work. (From the Bureau of Labor Statistics, 1999.)

accidents. Other costs associated with off-the-job accidents include absenteeism, lower production, and replacement worker costs. Historically, the focus has been on finding the least expensive medical benefit plans rather than ensuring preventive safety measures.

Absenteeism

Every year, absenteeism of American workers costs companies thousands of dollars. The total costs can range anywhere from 12 to 18 percent of a company's payroll, based on studies by national employee benefits consultants.[9] According to *HR Magazine*, absenteeism has increased 14.1 percent since 1992 and costs employers as much as $668 per employee.[10]

Another aspect of absenteeism that safety managers must consider is lost workdays, whether they result from on- or off-the-job injuries and illnesses. Since 1993, truck drivers, more than any other worker, have experienced the highest number of injuries and illnesses requiring time away from work.[11] As mentioned earlier, of the top ten occupations accounting for nearly one third of all injuries and illnesses requiring time away from work, truck drivers were number one. Moreover, the median number of lost workdays for all cases in 1999 was 6 days, with one fourth of these cases resulting in 21 or more days away from work.[5]

Lower Production

Another cost associated with off-the-job accidents is lower production through the loss of skilled labor. Although not easy to attribute directly to off-the-job injuries and illnesses, loss in production is a by-product of lost labor. Unplanned absences can delay or stop production for extended periods of time until replacement workers can be hired. Naturally, the exact loss in production depends on the specific job functions that the injured or ill employee performs.

Replacement Employees

Most employers do not realize that hiring replacements for employees who are injured or ill for extended time periods can be very costly. Many costs are associated with hiring replacement or temporary employees. Findings indicate that such indirect costs can be as high as 7 percent of payroll costs.[12]

Personnel or administrative costs constitute a large portion of the costs related to replacement employees. These expenses include advertisement of the position or contact costs for a temporary placement agency, as well as the costs of personnel resources used for screening, interviewing, and other human resources functions necessary in the hiring process.

Some employers also offer limited benefits to replacement or temporary employees. Legally required benefits typically account for 9 to 12 percent of total payroll, with workers' compensation alone accounting for 1.5 to 2.5 percent.[13] With indirect costs such as overtime pay for other employees filling in, hiring temporary

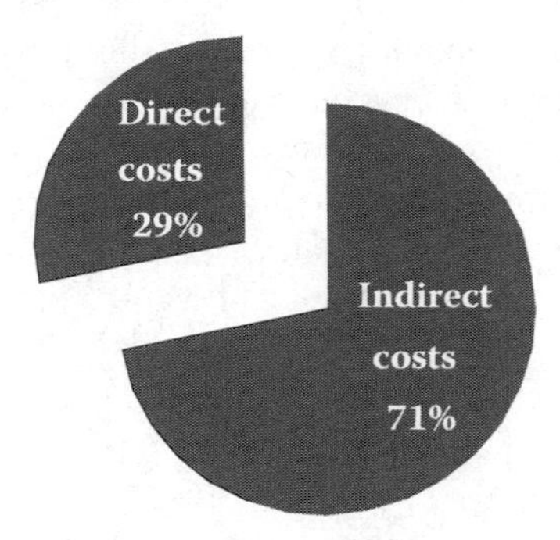

FIGURE 3.3 Indirect and direct costs of accidents.

employees, or using outside contractors costing up to 7 percent of payroll, accident prevention can save organizations money.

COSTS OF VEHICLE ACCIDENTS

As the leading cause of employee fatalities and a contributor to serious injuries, motor fleet vehicle accidents are costly. According to *National Highway Traffic Safety Facts 2000*, motor vehicle crashes cost more than $150.5 billion.[4] The average cost of a fleet accident, including direct and indirect costs, is $14,000, as shown in Figure 3.3. Consequently, economic reports estimate that organizations must produce $530 billion in profit to offset the costs of motor vehicle accidents and injuries.

Given that motor vehicle accidents cause numerous injuries and fatalities, it is essential that companies examine the various costs associated with them. The 1999 *Annual Statistical Bulletin* by the National Council on Compensation Insurance Inc. (NCCI) reported that the national average workers' compensation cost per workplace fatality was $167,847.[13] Other studies have shown that the total costs of such fatalities, including lost lifetime earnings and productivity as well as other direct and indirect costs, was $2.8 million per worker killed on the job.

Injuries and illnesses as a result of motor vehicle accidents are also very costly. The NCCI reported that the average cost involving permanent total disability for workers' compensation was $173,660, while the average cost for permanent partial disability was $21,093.[13]

CONCLUSION

In 2000, the National Center for Statistics and Analysis reported that an average of 115 people died each day, one every 12 minutes, in motor vehicle accidents.[4] The total deaths rose above 35,900, while the total injuries and illnesses requiring time away from work for truck drivers was reported at over 141,000. It is clear that American businesses are risking their competitive advantages, market share, and future abilities to remain viable by ignoring the impact of accidents resulting in deaths, injuries, and illnesses of their employees.

Reports indicate that an effective safety and health program focused on accident prevention not only reduces accidents but also reduces costs. Today's safety managers must go beyond the status quo and into accident prevention mode. This includes focusing attention and training on off-the-job injuries and illnesses and reinventing a safety culture that is second nature to all employees. Only when this level of safety consciousness is achieved will organizations be able to reap the benefits associated with it.

STUDY QUESTIONS

The following statements require True or False responses:

1. Organizations have no moral obligations to prevent accidents.
2. Vehicle accidents are the leading cause of fatalities in the United States.
3. Off-the-job accidents are a source of high expense for any organization.
4. Large trucks have a gross weight rating greater than 26,001 pounds
5. Since 1993, truck drivers have experienced more lost workdays than any other group of American employees.
6. The average cost of a vehicle accident is approximately $20,000.
7. The main types of fleet accidents are vehicle accidents, employee injury accidents, and off-the-job accidents.
8. Production is affected by absenteeism.
9. In 2009, over 35,900 people died in motor vehicle accidents.
10. Accident prevention is a way to reduce costs.

REFERENCES

1. *Workplace Injuries and Illnesses in 2000*, Bureau of Labor Statistics, Washington, D.C., 2001.
2. Fatality Analysis Reporting System (FARS), Web-based encyclopedia, http://www-fars.nhtsa.dot.gov.
3. Overview, *National Highway Traffic Safety Facts 2000*, National Center for Statistics and Analysis, Washington, D.C., 2000.
4. *Lost-Work Time Injuries and Illnesses: Characteristics and Resulting Time Away from Work, 1999*, Bureau of Labor Statistics, Washington, D.C., 2001.
5. Myre, J.C., Off-the-job safety, *Safety Times*, http://www.safetytimes.com.
6. Healthy people 2000 progress report for occupation safety and health, *Journal of Environmental Health* 59, 5, December 1996.
7. http://www.osha.gov/SLTC/safetyhealth_ecat/images/safpay1.gif.
8. Carrol, P., Strong return-to-work culture can cut costs of absenteeism, *Employee Benefit News* 14, 12, October 2000.
9. Martinex, M.N., Costs of absenteeism on the rise, *HR Magazine* 40, 11, November 1995.
10. *Number of Nonfatal Occupational Injuries and Illnesses Involving Days Away from Work by Selected Occupation and Industry Division*, Bureau of Labor Statistics, Washington, D.C., 1999.
11. Mitchell, R.W., *Employers Taking Steps to Reduce Costs of Absenteeism*, National Underwriter/Life and Health Financial Services 105, 15, April 9, 2002.

12. Hansen, F., Who gets hurt, and how much does it cost? *Compensation and Benefits Review* 29, 3, May/June 1997.
13. Transportation Safety Institute, http://www.tsi.dot.gov.
14. Sennewald, Charles A., *Effective Security Management*, 3rd ed., Butterworth-Heineman, Burlington, Massachusetts, 1998.
15. National Highway Traffic Safety Administration, 2007.
16. National Safety Council, *Injury Facts*. Itasca, IL: Author, 2011.

4 Special Considerations of a Small Fleet

In a small-fleet organization, the manager or owner may be responsible for planning and carrying out the fleet safety program. Even if the fleet consists of only a couple of cars, the responsibility remains to train employees in safe driving practices and techniques—yet the manager or owner may not know what goes into a fleet safety program. It is essential that the person in charge receive this information and implement it to the best of his or her ability because lives depend on it.

In a small-fleet organization, a vehicle operator is just as responsible for controlling and preventing accidents as the manager of a large fleet. Vehicle accidents can be especially detrimental to a small fleet because they can be costly. In an effort to prevent accidents, small-fleet managers must implement each part of a successful motor fleet safety program.

PASSENGER VANS

Today nine- to fifteen-passenger vans are used often by educational and private organizations for transporting youngsters to sports activities, camps, church services, etc. There have been a number of accidents with rollovers where passengers were ejected, causing some safety advocates to address these problems. Their concerns about these inherently dangerous vehicles in the late 1990s and early 2000s called for the National Transportation and Safety Board (NTSB) to research these crashes and to study the crashes involving fifteen-passenger vans. In 2010 the Federal Motor Carrier Safety Administration amended a regulation to improve safety with the van operators. Today, operation of nine- to fifteen-passenger vans is regulated by the federal government only for vehicles that operate in interstate commerce for direct compensation.

With the many problems and concerns with the nine- to fifteen-passenger vans, small organizations and church groups were advised to rent or purchase a small bus. Based on the many findings by the NTSB in 2005, a federal law made it illegal for school systems to purchase or lease a new fifteen-passenger van to transfer preschool and elementary school-aged children to or from school or an event related to school activities.

FLEET SAFETY PROGRAM ELEMENTS

The four elements of a fleet safety program[1] are as follows:

1. Setting management standards and policies
2. Recording accidents, injuries, and fleet safety program results
3. Selecting, training, and supervising employees
4. Encouraging and rewarding improved performance through awards, recognition, and the like

Motor fleets have an obligation to the public, their employees, and the community. New technologies and techniques encouraging drivers to perform safely must be applied in both small and large fleets. This will create a positive public image for a company and will better equip drivers to handle any adverse situation.

MAIN INTERESTS FOR FLEET OWNERS OR MANAGERS

Fleet owners or managers must commit to the safety of their fleets. This means conveying the safety message repeatedly and showing employees that owners and managers are committed to fleet safety. Setting up policies and procedures is just the start; this must be accompanied by the constant monitoring and tracking of performance. This essential step will tell how current safety practices compare with past ones, allowing managers to rate fleet safety performance. Monitoring is also important when rating drivers. If clear and accurate records are kept, managers can let drivers know exactly what they need to improve, and employees can be rewarded for doing an exceptional job.

Employees must know that anything less than total commitment to fleet safety is not acceptable. This will foster a positive safety mind-set of which everyone in the company will be aware.

CONSIDERATIONS IN EMPLOYEE SELECTION[6]

During the driver selection procedure, the interviewer will be able to uncover facts about the applicant and a decision can be made on whether to hire the individual. The information may come from different sources, especially past employment and recommendations from personal associates, medical examinations (drug tests), and personality screening tests. Employees can be expected to continue doing what they have done in the past. This is why employee selection is vital to the future success of a company. A company should hire individuals who have proven that they can be valuable assets. This allows the company to enjoy a greater profit margin and therefore have a better position in the marketplace.

According to the *Motor Fleet Safety Manual*,[2] a good driver must:

1. Avoid accidents.
2. Follow traffic regulations.
3. Perform pre- and posttrip inspections.
4. Avoid abrupt starts and stops.
5. Avoid schedule delays.
6. Avoid irritating the public.
7. Perform the nondriving parts of the job.
8. Find satisfaction in the job.
9. Get along with others.
10. Adapt to meet existing conditions.

Knowing what to look for is half of the battle. Selecting employees who will increase the company's profits could mean the difference between expanding or just surviving, especially for a small fleet.

Screening

The screening of employees starts before the interview. Employers may want to screen candidates before selecting people to call in for actual interviews. Screening can consist of a written test to identify tendencies that are either beneficial or negative in someone desiring a career operating motor vehicles. Another factor to consider is age. Studies have proven that drivers over age 25 and up to age 65 are as efficient as younger drivers but are safer workers, are less prone to job hop, and have lower rates of tardiness and absenteeism. Employers should also set minimum height and weight limitations based on the physical constraints of the driving compartment of the vehicle to be operated. There also appears to be a correlation between standardized, written, and behind-the-wheel test scores and satisfactory driving performance, as well as a direct correlation between satisfactory performance and the lifestyle of a potential employee.

Background Information Check

Each potential employee must complete an application for employment furnished by the motor carrier. According to 49 CFR 391.21,[3] the application must contain the following information:

1. The name and address of the employing motor carrier
2. The applicant's name, address, date of birth, and Social Security number
3. Addresses where the applicant has resided during the preceding 3 years
4. The date the application is submitted
5. All the information involving unexpired commercial driver's licenses issued to the applicant

6. The nature and extent of the applicant's experience in the operation of motor vehicles
7. A list of all motor vehicle accidents in which the applicant was involved during the 3 years preceding the date the application is submitted
8. A list of all violations of motor vehicle laws or ordinances (other than parking) of which the applicant was convicted during the 3 years preceding the date the application is submitted
9. A statement setting forth in detail the facts and circumstances of any denial, revocation, or suspension of any license, permit, or privilege to operate a motor vehicle or a statement that such activity has not occurred
10. A list of the names and addresses of the applicant's employers during the 3 years preceding the date the application is submitted
11. (For drivers applying to operate a commercial motor vehicle): Names and addresses of the applicant's employers during the 10 years preceding the date the application is submitted
12. The applicant's signature and the date when signed

Physical Examination

A complete physical and psychological examination should be administered by a physician to ensure that applicants are in good physical and mental health. Hiring healthy employees translates to huge medical savings for the enterprise.

Testing

Companies should administer tests to determine behind-the-wheel attitudes and abilities. These may include, but are not limited to, traffic and driving knowledge tests, driving skill tests, and attitude tests. These tests may suggest how well potential employees would perform the actual job.

Driver-Selection Process

In selecting new drivers for a small motor fleet company, management should ensure that the selection procedures properly evaluate the past history of all new hires. The fleet safety program should be designed to protect employees, to prevent and control accidents, and to increase the effectiveness of the drivers in order to reduce costs for the company. In the selection process, management must examine the general abilities and aptitudes of drivers, as they will be different from the demands required in specialized driving responsibilities (e.g., school bus drivers). The importance of selecting the correct driver when operating within a small fleet is to continue to maintain a very high level of safe driving.

Small motor fleet companies should incorporate complete applications, screening, background checks, physical examinations, and testing into the driver-selection process. There is less room for error in a small fleet than in larger ones. Every effort should be made to obtain good drivers who hold safety as a number one priority.

The steps just outlined should help small-fleet managers select highly qualified individuals to operate their motor vehicles.

PHILOSOPHY OF A SMALL-FLEET MANAGER

Small-fleet managers wear many hats. They should adopt the attitude that some training is better than none. Driver training, even in small fleets, can result in huge cost savings for the company. Fleet managers must train drivers on company procedures, inspections, defensive driving, and the use of seat belts. Each of these is vital to a smooth operation. If drivers do not know company procedures, how will they know how to react properly in emergency situations? Training drivers not only helps prevent incidents but also limits the extent of incidents that do take place.

Managers of small fleets may want to consider using outside agencies to train employees who drive motor vehicles. Safety managers may have skills unique to the operation but might not have those skills needed to train employees in the proper use of motor vehicles. In such cases, contracting with an outside agency may be beneficial to a company. An outside agency may have the skills, materials, and knowledge to train all employees without affecting the process of the business. This approach is often favorable to management and can strengthen the knowledge of motor vehicle drivers.

INITIAL TRAINING BRIEFS FOR NEW EMPLOYEES

Certain aspects of driving motor vehicles make it imperative that a company train employees driving these vehicles, especially when employees are new. Safety managers should never assume that new employees know the information that will keep them safe while operating a motor vehicle. It is the manager's job to train these employees to keep them, and others, safe. In this section, we look at some topics for training new employees.

Operating the Vehicle

New and transferred employees should always be trained in the safe and proper operation of the vehicles they are expected to drive. Each vehicle is unique, and each time employees change vehicles, they must be trained in the safe and proper operation of the new vehicle.

Inspecting the Vehicle Before Each Trip

Employees should be trained on how to complete a pretrip inspection. This is required by law for safety purposes. Tires, wheels, rims, lugs/nuts, steering, suspension system, exhaust system, and emergency equipment should all be inspected.[4] This inspection ensures that the vehicle is in proper working condition before it is driven off the lot.

Reporting Vehicle Defects

Whether operating in a small fleet or large one, the operator must have the ability to care for the vehicle he or she is driving. Some fleets that assign the same vehicle to the same driver each day have found that some of these drivers get more miles per gallon of fuel than other drivers. Because some vehicles require less work to repair, specifying the daily inspections and service needed creates a cooperative atmosphere among all the parties. This also ensures that driver complaints are investigated and repairs are made when necessary to make vehicles safe.

Training on the proper procedure used to report vehicle defects is vital to the success of a fleet safety and maintenance program. A vehicle that has defects needs to be taken out of service temporarily until maintenance personnel can fix the problem and restore the vehicle to proper working condition. By reporting vehicle defects in a timely manner, employees help ensure that vehicles are repaired and back on the road without major delays.

Filling Out Accident Reports

Proper training on the correct use and proper way to fill out accident reports is also critical. Employees should not be expected to know how to fill out these forms without training. Information included on the form should be as clear, concise, and correct as possible. Detail is very important, and all items should be completed. The accident report may be made available in court if injured parties file a lawsuit; therefore, the information must be correct and complete. Also, the accident report can play an important role in handling claims. An effective and sound accident-reporting system for a company is necessary to develop and maintain an outstanding accident-prevention program.

Using Seat Belts

According to the most recent report from the Federal Motor Carrier Safety Administration (FMCSA), safety seat belt use was up among commercial truck and bus drivers in 2010. FMCSA's safety belt survey found that safety belt use increased from 74 percent in 2009 to 78 percent in 2010. The researchers observed 26,830 commercial motor vehicle (CMV) drivers operating medium to heavy duty trucks and buses and 1,929 occupants at 989 roadside sites nationwide.

Also, safety belt use was 80 percent for CMV drivers and vehicle occupants in states where the safety belt law is primarily enforced versus 72 percent in states with secondary enforcement safety belt laws.

The use of seat belts has been proven to decrease significantly the number and severity of injuries sustained in vehicle accidents. This simple fact should be enough of a reason to mandate seat belt use for employees whose main job is operating motor vehicles. In addition, most states have laws requiring drivers and passengers to wear seat belts. By promoting seat belt use, a company presents a positive image to the public while protecting the lives of its employees.

Driving Defensively

Defensive driving means the driver must have both the desire and the ability to control accident-producing situations. It is, of course, quite essential that truck drivers stay alert to the movements of other drivers, but there is another important consideration to keep in mind. Concentration on driving large trucks must encompass both one's own driving and the driving of others. To concentrate exclusively on either one can be dangerous. A driver's positive, alert driving by watching himself or herself as well as other drivers is the best type of defensive driving. The defensive driver always follows the rules for better driving. He or she should always expect the unexpected and is therefore seldom surprised by the unsafe practices of other drivers. Also, defensive drivers accept responsibility for avoiding accidents and have a positive attitude that they can prevent them. Along with a good attitude, the defensive driver must demonstrate alertness, foresight, knowledge, judgment, and skill. All of these qualities can be developed in a training program and improved on through experience.

A direct correlation has been found between completion of a defensive driving course and reduced accident frequency. Developing and implementing a defensive driving course not only leads to reduced accidents but also looks good to the public and helps drivers understand the importance and usefulness of the material.

COMPONENTS OF A DRIVER SAFETY PROGRAM

The components of a driver safety program should encompass what drivers need to do their jobs well. Instructors should use the following categories[2] to rank focal points of the program:

1. *Vital* subject matter is absolutely essential to success on the job.
2. *Important* subject matter provides a basis for understanding the job.
3. *Helpful* material relates to the job and gives a broader base of understanding on which to build performance.
4. *Incidental* material is nice to know but not necessary for job performance.

Initial Training

Initial training time should be adequate to bring the knowledge and skills of the new driver up to the level needed to perform the job safely and properly. Money, time, and effort paid out in training costs will be offset by fewer accidents, lower maintenance costs, less absenteeism, a lower turnover rate, reduced supervisory burden, and improved public relations.

In-Service Training

Refresher training should be given annually or as needed. This should consist of classroom instruction where either initial training material is reviewed or updated material is presented.[5]

Remedial Training

Remedial training is designed for drivers who have had a certain number of accidents. Each remedial course should focus on the types of accidents the repeaters have had and should then incorporate discussion into the curriculum. A decrease in the number of accident repeaters has been a benefit of remedial training.

Ongoing Training

Education that exposes drivers to safety information and ideas should be incorporated on an ongoing basis. Some examples are safety posters, dash cards, bulletin boards, safety booklets, and driver letters. These items serve as constant reminders to act safely every day.

VEHICLE CONDITION

Safety and maintenance departments share the same goal: zero defects. The departments must work together to repair damaged vehicles prior to their release. It is vital to the success of the company that the safety and maintenance departments create conditions that allow repairs that make vehicles safe. This means working together for the best interest of the company.

Periodic Vehicle Inspections

These are annual inspections of a vehicle conducted by a qualified inspector. The inspection form should include the inspector's name, motor carrier, date of inspection, the vehicle inspected, the components inspected, and their conditions. These records must be kept in the vehicle.

Pretrip Inspection

Before drivers leave the lot, they should conduct a pretrip inspection. The U.S. Department of Transportation (DOT) requires that the components be checked and logged each time a vehicle leaves the lot. Some equipment included in a pretrip inspection are tires, wheels, rims, lugs/nuts, steering, suspension system, exhaust system, and emergency equipment. This inspection may take only 10 minutes, but it will give the driver peace of mind knowing that the vehicle is in good working order.

Posttrip Inspection

At the end of the workday, each driver is required to do a posttrip inspection. This step ensures the safety of the next driver. Both the pre- and posttrip inspections are similar, but the posttrip inspection must contain a statement either that there were no deficiencies or that gives an explanation of any deficiency found. The driver must also sign the report. If any deficiencies were detected, the mechanic and driver must sign off to indicate that they are satisfied with the repairs.

Vehicle inspections are tools used to eliminate unsafe vehicles from the road. An unsafe vehicle should never be permitted to leave the lot. This is the ultimate job of the safety manager: to ensure safe working conditions for employees. At times when productivity and the bottom line might encourage the operation of unsafe vehicles to save money, it is the job of the safety manager to step in and not permit this to occur. It takes the entire workforce committed to safety to do things in a way that will not be detrimental to employees or production.

MANAGEMENT'S ROLE IN A SUCCESSFUL SAFETY PROGRAM

Management should never be satisfied with knowing just enough to get by. Managers need to take advantage of available resources to further educate themselves as well as their employees. The Internet is a great source of information, especially websites for government agencies such as the Federal Motor Carrier Safety Administration (FMCSA), Federal Highway Administration (FHWA), and Occupational Safety and Health Administration (OSHA), as well as the American Trucking Association (ATA). The information found on these websites is current and touches on almost any topic relating to motor fleet safety. In addition, many journals and books dedicated to fleet safety are available. These may feature studies on motor vehicles and can provide valuable insight into potential problems.

Management should take the time to participate actively in the National Safety Council (NSC), local safety councils, and various trade associations. These groups provide information on all aspects of safety and will help managers and owners with questions relating to safety. These groups also provide current information that is industry or production specific and may give further insight into any problems.

Small-fleet managers should always set their goals high. In fact, small-fleet managers should set their performance to match the performance of a large vehicle fleet with a full-time safety staff. The main objective is to keep the public and employees free from vehicle incidents. Perfection is always the goal.

STUDY QUESTIONS

1. List the four main fleet safety program elements.
2. (True or False) Research shows that drivers over age 25 and up to 65 are as efficient as younger drivers and are safer employees.
3. Name four items that should be covered in initial training briefs for new employees.
4. Which categories should instructors consider when they are ranking topics on which to focus in a driver safety program?
5. What three inspections are required to be performed on a motor fleet vehicle?

REFERENCES

1. Della-Giustina, D., Elements of a fleet safety program, *Motor Fleet Safety*, West Virginia University, Morgantown, April 10, 2010.
2. Brodbeck, J.E., ed., *Motor Fleet Safety Manual*, 4th ed., National Safety Council, Itasca, Illinois, 88–108, 1996.
3. 15-passenger van alert. *Church Mutual*, February 8, 2008, http://www.churchmutual.com/index.php/choice/risk/page/rm-vanalert/id/35.
4. Code of Federal Regulations, Title 49, Part 391.21, U.S. Department of Labor, Washington, D.C.
5. Based on information from Tony Deligiannis.
6. Bird, F.E., Jr. and Germain, G.L., *Practical Loss Control Leadership*, 4th ed., International Loss Control Institute, Loganville, Georgia, 263, 1996.
7. *Fatality Analysis Reporting System* (FARS), web-based encyclopedia, http://www-fars.nhtsa.dot.gov, 2002.
8. Jabllon, R., NASA commuter van crash kills 3, injuries 7, *DailyLobo*, December 9, 2004, http:www.dailylobo.com/media/storage/paper344/news/2004/12/09/News/Nasa-Commuter,Van.Crash.Kills.3.Injuries.7-825961.shtml.
9. Overview, *National Highway Traffic Safety Facts 2001*, National Center for Statistics and Analysis, Washington, D.C., UMotor Fleet Safety and Security Management, 2nd Edition.S. Department of Transportation, 2002, publication no. DOT-HS-809-48.
10. *2000 Census of Fatal Occupational Injuries*, Bureau of Labor Statistics, 2001.
11. *Action Plan for 15 Passenger Van Safety*, National Highway Traffic Safety Administration, Washington, D.C., U.S. Department of Transportation, November 2004, http://www.nhtsa.dot.gov/problems/studies/15pass vans.
12. Newsome, T. and Meyers, D., *An Assessment of Tennessee Vans Driver Education Needs*, Knoxville, University of Tennessee, Center for Transportation Research, December 2008.
13. Pederson, S., personal communication, December 17, 2007.
14. Satterfield, J., Deadly van wreck: Trucker was too close, authorities say, *Knoxville News-Sentinel,* A1, A5, 1999.

5 Driver Selection

PREDICTING SAFE DRIVER QUALITIES

Predicting safe driving ability is not an exact science. Although there is no magic formula for identifying the perfect driver, this difficult and important process in a motor fleet safety program should still be included. Specific programs need to be developed and initiated so that future employees can be screened and allow the company to hire only the best, most qualified drivers, as well as the safest drivers available. Review the responsibilities of the job and decide which qualities are desired in an employee; then rate the importance of each of these qualities for the performance of the job's tasks. For example, here is a list of job qualities that might be rated when considering potential employees:[1]

- Education
- Experience
- Technical training
- Physical or visual demand
- Mental demand
- Responsibility for vehicle
- Responsibility for safety of cargo or passengers
- Responsibility for safety of pedestrians and other motorists
- Contacts with customers
- Responsibility for company funds
- Driving and working conditions
- Awareness of personal hazards
- Supervision received

SELF DISCIPLINE IN DRIVER SELECTION

Self-discipline plays an important role in driver selection. Many individuals are hired for their technical ability, and they must discipline themselves to keep their mind on driving. Since this is the case, it is important when screening drivers to look for self-discipline. Job factors such as responsibility for the vehicle, responsibility for the safety of cargo or passengers, responsibility for the safety of pedestrians and other motorists, responsibility for company funds, and awareness of personal hazards should have high ratings. These traits can be identified by reviewing past work history and performing reference checks.[1,2,3]

MANAGING DRIVER SELECTION

Driving defensively is an important element of the accident prevention program. At the 1970 National Safety Congress, Dr. Ray Martinez selected a number of facts important to driver selection. In the driver-selection process, defensive driver training is emphasized to avoid errors and how to avoid being trapped in accidents by the errors of others. The importance of driver training with the behind-the-wheel training phase is the most essential. The instructor who conducts the training program after the candidate is selected should have a well-organized program and specific objectives in the curriculum.

ABILITIES OF A GOOD DRIVER

When interviewing potential drivers, look for certain abilities associated with a good driver. Examples include the following:

- Drives well
- Performs the nondriving parts of the job well
- Finds satisfaction in the job
- Gets along with others
- Adapts to meet existing conditions

Driving Skills

When determining how well drivers are schooled in their trade, it is important to see how well they can drive. Look for certain factors while evaluating the skills necessary for the task at hand.

First, determine the driver's ability to avoid accidents. This may be the most important characteristic of the driver. No matter how well qualified a driver is in other areas of work, an accident repeater will always be a bad investment for the company.

Then look at the driver's ability to follow traffic regulations. This is also important because repeat regulation violators are just accidents waiting to happen. If such individuals are driving for a company, they can become a public relations liability.

A good driver is also one who cares for the vehicle. This is done through upkeep and maintenance. A fleet of clean vehicles can be a public relations boost to the company. An individual who cares for his or her vehicle and drives the same vehicle daily could show better gas mileage and in the long run save the company money.

The ability to meet scheduled deadlines is another major part of being a good driver. The company's schedules should be set so that a driver can stay in compliance with reasonable effort. When a driver gets behind, it throws off the whole schedule and can lead to accidents when drivers try to make up time.[1,2]

Performing the Nondriving Parts of the Job

Another factor is the driver's performance of the nondriving parts of the job. This could include many different things, such as making deliveries, keeping logs, servicing vehicles, caring for merchandise inside the vehicle, and even making sales. If drivers are not trained well in the nondriving parts of their jobs and are not comfortable with these duties, they can become a liability to the company.[1]

Job Satisfaction

It is important for individuals to be happy with their positions. If they are not, they can become a hardship to the company. They could perform their jobs poorly, which can cause the company losses. Employees must be looked at as investments; if they are not happy with their jobs, they might quit and then the company loses its investment. This is why it is important to keep qualified individuals satisfied with their positions. If employees are over- or underqualified, they might not find satisfaction in their jobs.[1]

Getting Along with Others

It is vital to employ workers who work well with others. Take into account applicants' personalities during the screening process, and then determine if they are a good fit with other employees. Always keep in mind that a company cannot hire just the quality parts of the candidate but must hire the whole person.[1]

Adapting to Meet Existing Conditions

Adaptability is critical because environments constantly change. Therefore, screen to see if potential employees can adapt to different situations.

PERSONNEL FACTS

For purposes of this chapter, the personnel facts that need to be considered are broken down into these categories:

- Age
- Sex
- Physical traits
- Intelligence
- Education

When considering these traits and screening for them, it will be important for the hiring manager to know the fair hiring guidelines for his or her state or province. The hiring manager should study these guidelines well so that he or she does not ask questions during the interview process that could get the company into trouble.

Age

Studies prove that drivers over 25 and up to 65 years of age are as efficient as young drivers, are safer workers, are less prone to job hop, and have low rates of tardiness and absenteeism.[2]

Sex

No basis exists for discriminating against women in motor vehicle driving jobs, given the comparative accident rates between the two sexes.[2]

Physical Traits

Minimum height and weight limitations based on the physical constraints of the driving compartment of the vehicle to be operated should be specified before beginning the hiring process.[2]

Intelligence

There appears to be no correlation between intelligence test results and satisfactory driving performance—except at the low and high ends of the scale.[2]

Education

In the present labor market, a high school education (or equivalent) is sufficient, especially for line-haul work.[2]

DRIVER SELECTION PROCEDURES

There are many procedures to use when hiring drivers. This section discusses eleven steps to consider incorporating when looking for qualified drivers. The steps are as follows:

1. Recruiting
2. Preliminary application
3. Application form
4. Credit-checking agencies
5. Check for operator's or chauffeur's license
6. National driver register
7. Employment interview
8. Physical examination
9. Reference check

10. Acceptance interview
11. Driving skills

Recruiting

Recruiting can play a huge role when looking for new drivers because a company does not always get the desired applicants when just posting the job. To hire the best qualified and safest drivers, a company may need help finding other applicants. Suggestions are through referrals, industry contacts, and driving and vocational schools.[1]

Preliminary Application

The preliminary application is simply a form for obtaining basic information from the applicant. In general, this is short and requests a condensed biography from the applicant.

Application Form

The application form should be developed specifically for a company and its needs. When a company creates its own application, it is important to take into consideration all fair hiring guidelines to which the company is subject. Alternatively, obtain a form from a personnel agency; anyone in the motor fleet industry could obtain and use this type of general form. Some sources for application forms are the American Trucking Association and agencies such as J. J. Keller & Associates.[1,2]

Credit-Checking Agencies

Because not everyone is honest, it is important to verify the information provided on applications, and it is a good idea to use a credit-checking agency. With credit checks, a hiring manager can see where applicants have lived in the past and sometimes what kind of lifestyle they may have led.[2]

Operator's or Chauffeur's License

This is a simple item that can be asked at any stage of the application or the interview, or both. Also, it is important to ask what endorsements applicants may have on their licenses.[1,2]

National Driver Register

The National Driver Register (NDR) is a central repository of information on individuals whose privilege to drive has been revoked, suspended, cancelled, or denied or who have been convicted of serious traffic-related offenses. The records maintained at the NDR consist of identification information including name, date of birth, gender, driver's license number, and reporting state. All of the substantive information, the reason for the suspension or conviction, and associated dates resides in the reporting state.

State driver licensing officials query the NDR to determine if an individual's license or privilege has been withdrawn by any other state. Other authorized users have access to the NDR for transportation safety purposes. All fifty states and the District of Columbia participate in the NDR. The system is also referred to as the Problem Driver Pointer System (PDPS).

By checking with the NDR, a hiring manager is able to verify information provided by candidates on their applications. Also, the great amounts of information catalogued on the Web make it almost impossible for drivers to obtain licenses in different states. This check will serve as another verification of the facts that the applicants have already provided.[2]

Employment Interview

This is the most important step in the employee-selection process, so the interviewer must be just as prepared as the applicant. Review the application carefully beforehand and use the interview time to expand on the information from the application as well as to verify it. This is also a good time to see what kind of person the applicant is. The interviewer should get an idea of how well the individual will be able to work with other current employees. At this point, the hiring process is far from over, so it will be important not to mislead the applicant into thinking that he or she has the job.[1]

Physical Examination

If the applicant has made it this far, a physician should determine whether the applicant is fit for the job. The company must have a good relationship with a physician who is familiar with the kind of work that the applicant will be doing if hired. This is when the initial screening for alcohol and drugs should be held. It is critical that the doctor sign off after the examination. This way, the physician is liable rather than the company if the interviewer missed something in the examination process.[1,2,3]

Reference Check

The world we live in is not perfect, which means that hiring managers must check that applicants have not lied to them. The interviewer must do this to see if applicants really worked for employers they listed on their applications and if their job duties were as they reported. The best way to do this is via the telephone. Take care of this step quickly because everyone's time is valuable. Have questions prepared so that during any conversation with the applicant's former supervisor the phone call can go quickly. Another way to get this information is to send a form letter to the listed companies, but not all companies will respond; some will fail to return the letters and others will be hesitant to sign their name to the letter (which could be interpreted to mean that what they had to say was not good). Another method is to make a personal visit to the applicant's former company. This may be the hardest way to do it because of the cost involved with the process.[1,2]

Acceptance Interview

This is the time that the hiring manager brings the applicant in and to see if he or she is still interested in and enthusiastic about the job. This is a critical time because it will be the last chance for the interviewer to change his or her mind about the applicant. At this time, the interviewer will need to discuss salary, benefits, and hours with the applicant. After that, the interviewer can present the formal job offer and finalize the acceptance letter.[1]

Driving Skills

At this point, the driver will go through initial or refresher training. This can be done off-site through another company or by a company's in-house training staff. Another option is to use one of the many trucking rodeos held around the country.[2] Rodeos held on driving ranges help support and develop drivers' skills.

ELEMENTS OF PRESELECTION SCREENING RELATED TO DRIVING

For the purpose of this chapter, we will concentrate on four main elements of preselection screening:

- Driving experience
- Past driving record
- Previous driver instruction
- Physical fitness

Driving Experience

When looking at this area, consider the type of vehicle, length of time, type of operator's license, and type of driving. With all these factors in mind, a hiring manager should find it easy to decide what is most important for the open position.[2]

Past Driving Record

Always check on the applicant's previous traffic violations and accident history. The hiring manager will want to get the applicant's professional and personal information.[2]

An *accident repeater* is a person who, in the same or in different situations, seems to continue having the same type of accident. A person with few or no accidents may indicate self-discipline.

FMCSR's Responsibility When Checking Driving Records

Based on the Federal Motor Carrier Safety Regulations (FMCSR-391.23), it is the responsibility of the employing carrier to check into a driver's driving record for the past 3 years from each state in which the driver held a motor vehicle operator's license or permit during those 3 years.[4] The driving record of each driver that is employed must be reviewed at least once each year. A copy of all correspondence obtained must be placed in the driver's personnel file within a 30-day period of the date the new driver commences employment. Also, an investigation of the driver's safety performance history (if any) with Department of Transportation–regulated employers during the preceding 3 years must be conducted. The investigation may consist of personal interviews, letters, telephone interviews, or other methods of checking out the driver's history, if deemed appropriate.

Previous Driver Instruction

The hiring manager should ask about previous driver instruction so that he or she will know what type of investment will have to be made in training the new employee. All classroom and behind-the-wheel instruction should be noted. Find out how much instruction was given, who provided the instruction, and what type of vehicle was involved in the instruction.[2]

Physical Fitness

At this point in the process, set minimum standards of physical fitness for the job. Then decide if a physical examination is necessary and determine the frequency rate for any periodic exams.[2]

TESTS TO DETERMINE BEHIND-THE-WHEEL ATTITUDES AND ABILITIES

It is a good idea to incorporate more than one test to find out about a person. Three different types of tests can be administered:

- Traffic and driving knowledge tests
- Driving skill tests
- Attitude tests

These tests should give a good idea of the driver's total abilities. The written, driving, and psychological tests will let a company know more about the driver's total skill set.[2,3] Psychological testing analyzes the sum of a person's actions, traits, attitudes, and thoughts.

STUDY QUESTIONS

1. (True or False) A driver who has repeated traffic violations will likely end up with repeated traffic accidents.
2. What or who should always be consulted before you design any questions for your application or conduct an interview?
   a. Interviewing handbook
   b. Fair hiring guidelines
   c. The person applying
   d. Your mother
3. (True or False) The interview is the most important step in the screening process.
4. How can the company profit from a driver's care of the vehicle?
   a. Improved gas mileage
   b. Improved company public relations
   c. Lower upkeep costs on the vehicle
   d. All of the above
5. Which is a personnel fact that should be considered when hiring?
   a. Marital status
   b. Sex
   c. Race
   d. Age

REFERENCES

1. Brodbeck, J.E., ed., *Motor Fleet Safety Manual*, 5th ed., National Safety Council, Itasca, Illinois, 2010.
2. Della-Giustina, D., *Safety and Environmental Management Workshop for Safety Policies and Practices*, West Virginia University, Morgantown, 2002.
3. Code of Federal Regulations, Title 49, Part 391, U.S. Department of Labor.
4. *Federal Motor Carrier Safety Regulations Handbook*, J. J. Keller & Associates, Inc., Neenah, Wisconsin, 2009.

6 Driver Training and Instruction

BENEFITS OF DRIVER TRAINING

Driver training gives a company and its employees many useful benefits. One of the benefits gained involves the insurance program. The fewer claims made by a company, the lower the insurance premium.[1] With a lower insurance premium, the company can use the extra money to buy new equipment, for example. Another benefit relates to the actual equipment. The better care provided to equipment, the lower the maintenance costs, which in turn extends the life of the equipment. The better maintenance given to a vehicle, the less downtime for that particular vehicle, which saves a company from possibly having to rent a vehicle like one it already owns.[1] Training drivers can improve the efficiency of the company because when employees know what is expected of them and the company shows that it cares, employees are more likely to do their work as efficiently as possible. For the past 40 years, the training of truck drivers has been an area of adult education service to the traffic safety movement. More employment exists, particularly in metropolitan areas where there are a large numbers of drivers available who could profit from this experience. This has been a very successful operation for a number of school districts because of a shortage of part-time school bus operators. The instruction for fleet and school bus drivers was implemented primarily by qualified commercial fleet supervisors who already owned the heavy equipment necessary to conduct the program for industry and private truck companies.

Driver training is a process that cannot be the same for everyone because some drivers may need more training and some less training. Training can be separated into four steps.[2] The first step is called the *initial training*. This stage deals with the knowledge and skill required of a driver to perform the job correctly. The next step is the *refresher training*. This stage consists of a few days of classroom instruction used to update drivers on news, rules, and new equipment. Refresher training is held once a year or as needed, depending on the introduction of important rules and regulations that drivers must know about in order to perform their jobs safely. The third step is called *remedial training*. Remedial training is mainly used for drivers who have had accidents while driving. This stage is designed to inform drivers of what they are doing wrong and to identify any lack of knowledge they may have so they can avoid accidents. The last stage, considered the most important, is *ongoing training*, which is necessary to keep drivers on track and keep them from falling into any unsafe habits. Ongoing training includes classroom training, safety posters, bulletins, driver handbooks, and letters to the drivers to inform them of their performance.

BENEFITS OF USING A TRAINING ROOM

Training rooms should have blackboards, projectors, and any other aids that an instructor finds useful for students. The room can have many uses: Meetings and conferences can be held there, and it can serve as a display area for important material that companies want to make available to drivers (for instance, they can read up on the latest news in the trucking industry).

SYSTEMATIC VISUAL HABITS IN DRIVING

While driving, a person must always pay attention to what he or she is doing as well as be aware of the surroundings. According to the *Florida Commercial Driver Handbook*,[3] drivers should follow the defensive driving technique called the Smith System:

1. Aim high in steering.
2. Take in the whole picture.
3. Always keep eyes moving.
4. Leave yourself an out.
5. Make sure other drivers can see you.

Defensive driving involves doing all you can to prevent crashes. Defensive driving can be shortened to three steps: Look for possible danger, decide what can be done to prevent an accident, and act quickly.

The IPDE process can also be used while operating a vehicle. The IPDE process is described in the *Emergency Vehicle Operator's Student Guide*. This process lists four steps that will help drivers avoid accidents:

1. *Identify* any hazards that may cause an accident.
2. *Predict* where an incident may occur.
3. *Decide* what action to take in order to avoid the hazard.
4. *Execute* your decision on what corrective action you will take.

Be sure to incorporate these ideas into any driver training program.

PROCEDURES TO FOLLOW IN CASE OF AN ACCIDENT

When an accident occurs, a driver must follow certain procedures to protect others. The first step is to secure the scene. Securing the scene means stopping the vehicle immediately and not moving any other vehicles involved unless there is a chance of fire. A driver must then turn on the four-way flashers, put the vehicle in the lowest gear, and set the parking brake before leaving the cab. Once out of the cab, the driver needs to remain calm and find out if anyone is injured. It is important to set out warning devices to warn other drivers that an accident has occurred. Some examples of warning devices are triangles, flares, fuses, reflectors, and other acceptable reflective or luminescent equipment.

The next step is to notify the proper authorities. This involves calling the police, the driver's company, and, if necessary, medical assistance. If carrying hazardous materials, the driver must notify the Chemical Transportation Emergency Center (CHEMTREC) to arrange for cleanup of any possible leakage.[3]

The third step is to document the accident. A driver should carry courtesy information cards in order to collect information from eyewitnesses. The driver should ask an eyewitness to fill out his or her name, address, and telephone number so the person can be contacted at a later time. If possible, the witness should also write a description of what he or she saw.

A driver is also supposed to fill out a preliminary accident report while at the scene once everything is secured. This report is used to gather all information about everyone involved in the accident and any eyewitnesses. A driver must record the time, place, and road and weather conditions and provide a short description of what happened, which can be discussed later in case of lawsuits and other actions. If a camera is available, a driver should take photographs of the scene to give the insurance company an idea of what took place when it reviews data at a later time. A driver must not sign anything or make any statements unless it is to the police, his or her company, or the insurance company.

Again, be sure to include this information in a driver training program.

METHODS OF TEACHING DRIVER TRAINING

When teaching driver training, an instructor must cover all the rules and regulations for driving. The first step is to teach the rules to follow while on the road. Even if a driver follows the rules all the time while driving, it does not mean that he or she will never be involved in an accident; there are many other drivers on the road who do not always follow the rules and could cause an accident. The instructor should use videotapes or DVDs that show defensive driving techniques and explain what to do in situations that occur on the road and what to do in the case of an accident. The instructor should supply drivers with important reading materials (or tell the class where to locate reading material). Once the drivers have reviewed all the information from the instructor, they must be evaluated in two areas: a written examination and an actual behind-the-wheel road test. The road test should be given only after a driver has passed the written exam and the instructor has given a full demonstration of what is expected on the road test.

ADDITIONAL TOPICS TO COVER IN TRAINING

In this section, we discuss several topics that a good driver training program must incorporate.

Factors That Affect Stopping

When preparing to stop, the driver should keep in mind three elements that factor into the distance it will take to stop the vehicle. The first is *perception distance*, which is the distance a vehicle travels from the time a driver sees the hazard until a driver's brain recognizes it. The perception time for an average driver is 3/4 of a second. In that 3/4 of a second, a vehicle can travel 60 feet at 55 mph. The second element is *reaction distance*, which is the time it takes the driver's brain to move his or her foot to the brake. Once again, this is 3/4 of a second, which means that a vehicle traveling at 55 mph will add another 60 feet to the amount of distance it takes to stop a vehicle. The third element is *braking distance*, which is how far it takes the vehicle to stop once the brakes are applied. At 55 mph a heavily loaded vehicle will take about 170 feet to stop. The total stopping distance is about 290 feet. A vehicle's stopping ability can be affected by many factors, including tire condition, braking power, traction, load of the vehicle, the speed of the vehicle, road surface, and weather conditions.

Traction is the friction between the tires and road surface. Sufficient traction allows a vehicle to speed up, slow down, and make any maneuvers safely. The friction keeps the tire from sliding on the road surface. Traction depends on the condition of the tires and whether the road surface is dry or slippery. The more slippery the road surface, the longer it will take to stop the vehicle, even if the tires are in good shape. A heavy load affects the stopping ability of a vehicle because the heavier the load, the more that load pushes the truck forward. This is based on the principle of inertia, which keeps an object pushing while in motion. The speed of a vehicle also affects braking ability; the faster a vehicle is moving, the longer the braking distance.

Methods of Judging Good Brakes

When identifying weak brakes, inspectors use a performance-based brake test (PBBT). PBBTs can assess the braking capability of a vehicle through a quantitative controlled measure of both individual braking and overall vehicle performance. PBBTs are beneficial to law enforcement and the freight community because they provide a standard measure of a vehicle's braking performance. Brakes are measured in terms of forces and weights, as described here:

1. A minimum force at a given air pressure for pneumatically braked vehicles developed by the National Highway Traffic Safety Administration's Vehicle Research Test Center.
2. A minimum ratio of brake force balance across an axle of 0.65 or better for any vehicle or brake type.

3. A minimum brake force as a function of wheel load. On the steering axle, a ratio of 0.25 is recommended; on a nonsteering axle, a ratio of 0.35 is recommended. These ratios are calculated using the brake force and wheel load (bf/wl).

The following criteria determine what places a vehicle out of service (OOS):

1. A vehicle will be placed out of service if 20 percent or more of its brakes are defective according to the PBBT criteria.
2. The vehicle will also be placed OOS if it cannot stop within a given distance.
3. A vehicle must have the capability of being stable by staying in its designated lane.

Distractions for Fleet Drivers

According to the National Safety Council's publication *Injury Facts*, during 2009 more than 5,500 people were killed as the result of distracted driving accidents. Approximately 25 percent of injury crashes were based on distracted driving that resulted in over a million injuries. These rates did drop somewhat from 2008, but distractions continue to injure and kill people in needlessly numbers.[5]

Some of the distractions while driving include talking or texting on cell phones, eating, drinking, grooming, and talking to other passengers while driving. Regular monitoring will show that many of the accidents are related to distracted driving, especially cell phone use. On January 3, 2012, FMCSA passed a final rule banning commercial motor vehicle drivers from using handheld cell phones while driving on interstate highways.

Also, distraction happens when a driver is slow to recognize a potential hazard because something inside or outside the vehicle draws the driver's attention away from the road. Approximately 25 percent of crashes are distraction related.

Distracted Drivers Test

Recently the National Broadcasting Company (NBC) created a quiz for their current events program *Dateline*. This program was developed by leading scientists interested in analyzing driver distractions. This quiz was not intended to be scientifically valid but was to serve as a guide to determine the risk factor of everyday drivers. It was intended to make individual drivers more aware of any dangerous driving habits that they may have developed over the years. Based on this quiz, a number of motor carrier fleet safety directors became interested in adopting and implementing the approach as part of their training program.

Motor fleet companies should develop and implement a cell phone policy that requires all drivers to avoid using cell phone or other electronic devices while driving. Drivers could allow voice mail to handle their calls, but if the driver must use a cell phone he or she should pull off the road and stop the vehicle. In the long run, reducing driving distractions by the driver will reduce the financial obligations, whether the company is self-insured or with an insurance carrier. This would provide a strong financial incentive to reduce accidents with significant success and a strict driving policy.

A sophisticated distracted driving policy for all employees states to the employees that their employer actively cares about their safety when behind the wheel.

Safety Considerations When Starting a Truck

Every type of equipment must be inspected for use. The inspection is designed to ensure safety and to prolong the life of the equipment. When conducting an inspection on a vehicle, the driver must first make sure that the wheels are chocked and the parking brakes are set. The next step is to inspect under the hood of the vehicle. Here a driver checks the levels of the oil, coolant, windshield washer fluid, transmission fluid, power steering fluid, and hydraulic fluid. While performing the inspection, the driver must also note the conditions of the hoses and belts and look for any cracks in electrical equipment. The driver must look underneath the vehicle to check for any leak spots on the pavement. Once the driver has finished the inspection, he or she must start the engine and listen for any unusual noises before attempting to move the vehicle.

Important Safety Rules in Preventive Maintenance

A company that practices preventive maintenance on its vehicles is a company that wishes to keep its vehicles in safe working order. Another objective of doing preventive maintenance on a vehicle is to reduce accidents that are caused by vehicle defects. When performing preventive maintenance, the driver and mechanic must check several items. For instance, a driver must ensure that the vehicle is in safe working condition. The vehicle must not be abused unnecessarily in any way. The driver must always do a pretrip and a posttrip inspection of the vehicle and note any problems that have been detected. If any severe deficiencies are noticed at any time, the driver should not move the vehicle.

Preventive maintenance also involves checking the brake system, including the brake shoes and the air lines to the brakes. Tires should be checked for inadequate inflation, too much wear, and any deterioration that could cause an accident. The steering system must be checked because too much play can cause a driver to be unable to turn the vehicle appropriately. Another important section of the vehicle that must be checked is the trailer coupling connection, which consists of the fifth wheel and adjustable axles. The fifth wheel is where the trailer is connected to the cab, and it must be secure to prevent the trailer from becoming detached. The adjustable axles are important because if they are not locked securely, the trailer can slide right off them and cause a major accident.

For safety purposes, all of a vehicle's lights must be operational so that other drivers can see the vehicle. The vehicle must have reflectors in case of heavy fog or partial failure of the lighting system.

Preventive maintenance is used for both vehicle life and the safety of all people on the road.

FLEET SKILL TESTS

The purpose of having a driver perform a skill test is to evaluate how that driver handles a vehicle. The skill test consists of performing a set of exercises during a road test. The test can also be done at the training site so that the instructor can observe a driver performing maneuvers without the chance of serious property damage. Driving skills can be determined to some degree by evaluating a driver's past driving experience and a review of his or her commercial driver's license (CDL) endorsements. According to the National Safety Council's *Motor Fleet Safety Manual*, most fleet safety directors support two types of driving tests. The first one involves the exercises on a driving range where they are constantly tested on a variety of different skills. The second skill is where the drivers are exposed to different kinds of actual traffic conditions. Some of these activities take place in heavy city traffic, light city traffic, secondary roads, and expressway driving to identify whether the drivers can handle the different traffic patterns. These skill tests show the applicant's skill in handling certain types of vehicles as well as the conditions under which they might be expected to operate.

The exercises usually feature six areas in which a driver must perform successfully. The following areas are discussed in the *Maryland Commercial Driver's License Manual*:

1. Backward serpentine: A driver is required to back through a set of three cones without touching any of them.
2. Alley docking: Alley docking requires a vehicle to be backed down an alley to a certain point where cones are placed, and the driver must get as close to them as possible.
3. Forward stop: In this exercise the driver must drive a vehicle between two sets of cones and get as close to the end of them as possible without crossing an imaginary line between the two sets of cones.
4. Straight-line backing: The driver must back through two sets of cones without running over any of the cones.
5. Parallel parking: The first type of parallel parking is along the driver's side. This type of parking is along the empty space on the left side of the vehicle. The space is long enough for the truck to fit and has cones placed at the end, signifying an object behind the space where the truck is trying to park. The second type of parallel parking is the conventional type, in which the space is on the right side. The setup is the same as the driver's side, and the object of both is to avoid hitting the cones behind the space where the driver is attempting to park.
6. Right turn: The object of this exercise is to ensure that a driver can make a turn properly without cutting it too short and running up on a curb or an object that may be located near an intersection.

When employees drive defensively and skillfully, they have both the desire and the ability to control accident-producing situations.

STUDY QUESTIONS

1. What are the benefits of driver training?
2. Name the four steps of driver training.
3. What makes up the IPDE process?
4. What three elements factor into stopping a vehicle?
5. What does PBBT stand for? Give the definition of PBBT.
6. What three criteria will render a vehicle out of service?
7. What is the purpose of performing preventive maintenance on a vehicle?
8. What is a fifth wheel?
9. Name the six exercises a driver should perform on a road test.
10. (True or False) Preventive maintenance on all trucks should be an ongoing program.
11. The Distracted Drivers Test was adopted by what company?
12. When carrying hazardous materials, the driver must notify what agency to arrange for cleanup of any possible leakage?

REFERENCES

1. Brodbeck, J.E., ed., *Motor Fleet Safety Manual*, 4th ed., National Safety Council, Itasca, Illinois, 1996.
2. Peterson, D., *Techniques of Safety Management*, American Society of Safety Engineers, Des Plaines, Illinois, 1998.
3. Kujat, J.D., *Fleet Safety Made Easy*, Government Institutes, ABS Group, Rockville, Maryland, 2001.
4. Meyer, M.D., and E.J. Miller, *Urban Transportation Planning: A Decision-Oriented Approac*h, McGraw-Hill, New York, 1984.
5. National Safety Council. *Injury Facts*. Itasca, IL, 2010.

7 Driver Supervision

RESPONSIBILITIES OF DRIVER SUPERVISORS

The driver's primary responsibility is to head off accidents before they occur. The driver supervisor must be capable of identifying potential incidents that may be costly to the employees and the company. To reduce potential incidents, the supervisor must be able to answer these three questions:

1. What are the most frequent errors?
2. How can these errors be identified?
3. How can errors be prevented?

In an article from *Fleet Owner*, fleet manager Ron Uriah states, "Our safety message is strong. It starts with the employment interview and we look for people with strong work ethics and good attitudes. We only hire experienced drivers and their first 90 days is a probationary period."[1] Often in smaller companies the fleet manager or owner is primarily responsible for assigning delivery and pickup routes. In larger companies, several persons may share the supervisory responsibilities. Both the supervisor's and management's support are key to an effective motor fleet safety program.

According to the *Motor Fleet Safety Manual*, several other measures can be taken to supervise drivers.[2] For example, safety road patrols, insurance companies, or contract services can make on-the-spot observations of driving performance and report back to company headquarters. Sometimes the public can provide valuable feedback as well. Using clearly posted signs or decals on vehicles can also aid in performance evaluation. Programs such as How's My Driving? guarantee "at least 10% reduction of accidents in the first year or services are free."[2]

The use of incentives for drivers may also help reduce incidents in the motor fleet. Awards for "no incidents" or "miles driven incident free" are reward programs that sometimes motivate drivers to maintain safe driving behaviors in the short as well as the long term. Having meaningful rewards for desired performance may help ensure a risk-free environment. An article from the *Journal of Applied Psychology* indicates that it is possible that supervisors—given their ability to provide reinforcement through praise, performance appraisal, and reward power—may play a role in creating an environment in which drivers perceive a strong safety climate. The same article also indicates that the safety attitude of fleet managers and supervisors may be one of the largest predictors of positive safety culture, which could lead to a reduction in motor vehicle accidents. "At the individual level, drivers' perceptions of the safety values held by their supervisor and their fleet manager, along with drivers' own driving self-efficacy and safety attitudes, were identified as proximal predictors of driver safety motivation."[7]

DEPARTMENT OF TRANSPORTATION

The U.S. Department of Transportation (DOT) was established on April 1, 1967, by an act of Congress on October 15, 1966. The mission of the department is to "serve the United States by ensuring a fast, safe, efficient, accessible and convenient transportation system that meets our vital national interests and enhances the quality of life of the American people, today and into the future."[10]

FEDERAL MOTOR CARRIER SAFETY ADMINISTRATION

The Federal Motor Carrier Safety Administration (FMCSA) was established within the Department of Transportation on January 1, 2000, pursuant to the Motor Carrier Safety Improvement Act of 1999 (Public Law No.106-159, 113 Stat. 1748, December 9, 1999). It was formerly a part of the Federal Highway Administration. The FMCSA's primary mission is to prevent commercial motor vehicle–related fatalities and injuries. Administration activities contribute to ensuring safety in motor carrier operations by enforcing safety regulations, targeting high-risk carriers and commercial motor vehicle (CMV) drivers, improving safety information systems and commercial motor vehicle technologies, strengthening commercial motor vehicle equipment and operating standards, and increasing safety awareness.

To accomplish these activities, the administration works with federal, state, and local enforcement agencies, the motor carrier industry, and labor safety interest groups, among others.

Another source of information can be obtained through the Commercial Vehicle Safety Alliance. To achieve a rating of "satisfactory" in the vehicle category, carriers must have no critical or acute violations, and their out-of-service rate must be less than 34 percent.

Compliance with FMCSA

We need to be aware of the DOT regulations for compliance. *Commercial motor vehicle* means any self-propelled or towed motor vehicle used on a highway in interstate commerce to transport passengers or property when the vehicle:

1. Has a gross vehicle weight rating (GVWR) or gross combination weight rating, or gross vehicle weight or gross combination weight, of 4,536 kg (10,001 lbs.) or more, whichever is greater; or
2. Is designed or used to transport more than eight passengers (including the driver) for compensation; or
3. Is designed or used to transport more than fifteen passengers, including the driver, and is not used to transport passengers for compensation; or involved in interstate commerce with vehicles with a combined gross vehicle weight rating (CGVWR) of over 10,001 lbs.

Commercial Motor Vehicles

If the company fleets have the following weight rating, then they are considered to be a commercial motor vehicle:

- Pickup truck GVWR = 9500 lbs.
- Trailer GVWR = 13,000 lbs.
- CGVWR = 22,500 lbs.

Driver Requirements/Responsibilities

According to FMCSA Parts 391.11 and 391.13, a driver must meet the following requirements and responsibilities:[9]

- Be at least 21 years of age
- Speak and read English well enough to converse with the general public, understand highway traffic signals, respond to official questions, and be able to make legible entries on reports and records
- Be able to drive the vehicle safely
- Be in good health and physically able to perform all duties of a driver
- Possess a valid medical certificate
- Have only one valid commercial motor vehicle operator's license
- Provide an employing motor carrier with a list of all motor vehicle violations or a signed statement that the driver has not been convicted of any motor vehicle violations during the past 12 months; a disqualified driver must not; be allowed to drive a commercial motor vehicle for any reason
- Qualified to drive a CMV
- Pass a driver's road test or equivalent
- Know how to safely load and properly block, brace, and secure the cargo

Records Needed to Be Kept on File

- Proof of insurance
- Driver qualification files
- Driver's application for employment
- Inquiry to previous employers—3 years
- Annual review of driving record
- Annual driver's certification of violations
- Copy of driver's license
- Medical examiners certificate
- Federal Motor Carrier Safety Regulations (FMCSA) handbook acknowledgement
- Company vehicle policy acknowledgement
- Maintenance records

Employer Responsibility

No employer shall knowingly allow, require, permit, or authorize a disqualified driver to drive a CMV. The period of time that a driver must be disqualified depends on the offense and the type of vehicle the driver was operating at the time of the violation.

FMCSA Safety Review and Rating

The Motor Carrier Safety Measurement System (SMS) is a tool used by FMCSA and some states to evaluate a carrier's safety performance. The online safety measurement system can be accessed at http://ai.fmcsa.dot.gov/SafeStat/CarrierOverview.asp?ais=&dot=1806793&WhichForm=.

Along with setting regulations, the FMCSA has recently instituted the Compliance, Safety, and Accountability (CSA) program, which monitors the safety performance of motor carriers through Behavior Analysis and Safety Improvement Categories (BASIC) sources. The CSA website states, "SMS uses a motor carrier's data from roadside inspections, including all safety-based violations, State-reported crashes, and the Federal motor carrier census to quantify performance in the following Behavior Analysis and Safety Improvement Categories (BASICs)."[8] The categories rated are Unsafe Driving, Fatigued Driving (Hours of Service), Driver Fitness, Controlled Substances/Alcohol, Vehicle Maintenance, Cargo Related, and Crash Indicator.

Rules for Drivers' Hours of Service

According to the FMCSA, hours of service (HOS) are the number of hours a CMV driver (who drives a vehicle that has a GVWR of 10,001 lbs. or more) is allowed to drive each day and week. The regulations do not restrict the hours a driver can work, but they do affect the amount of time the driver can drive. HOS is based on the driver's time on duty, not the actual driving time. Time on duty continues to count toward the available hours to drive.

Hours of Service

Driving Time: Maximum of 11 hours
Off-Duty Time: 10 hours
On-Duty Time: No CMV driving beyond 14 consecutive hours. A driver may work but cannot drive a CMV after 14 hours on duty
Duty Time: 60 hours in 7 days. A driver may work but cannot drive a CMV after 60 hours on duty in 7 days.
On-Duty Restart: 60-hour clock goes to zero after 34 continuous hours off duty and driver restarts the HOS.

Driver's Daily Log

CMV drivers *must individually record* their duty status on a driver's daily log (see Figure 7.1).

- No exemptions are allowed.
- Driver must complete a driver log for each day of the month. This includes rest days, vacation, and days the driver does not drive a CMV.
- Drivers maintain previous 8 days' logs.

Requirements for Vehicle Maintenance

- Every carrier shall systematically inspect, repair, and maintain all commercial motor vehicles under its control.
- Motor carriers must maintain the following information for every vehicle they have controlled for 30 days or more:
  - Identifying information, including company number, make, serial number, year, and tire size
  - Schedule of inspections to be performed, including type and due date
- Inspection, repair, and maintenance records

Roadside Inspection Reports

- Any driver who receives a roadside inspection report must deliver it to the motor carrier (FMCSA 396.9).
- Within 15 days after the inspection, the carrier must sign the completed roadside inspection report to certify that all violations have been corrected

Posttrip Inspection Report

Every carrier must require its drivers to prepare a daily written posttrip inspection report at the end of each driving day (FMCSA 396.11).

Periodic Inspection

Every commercial vehicle, including each segment of a combination vehicle requires periodic inspection that must be performed at least once every 12 months (FMCSA 296.17).

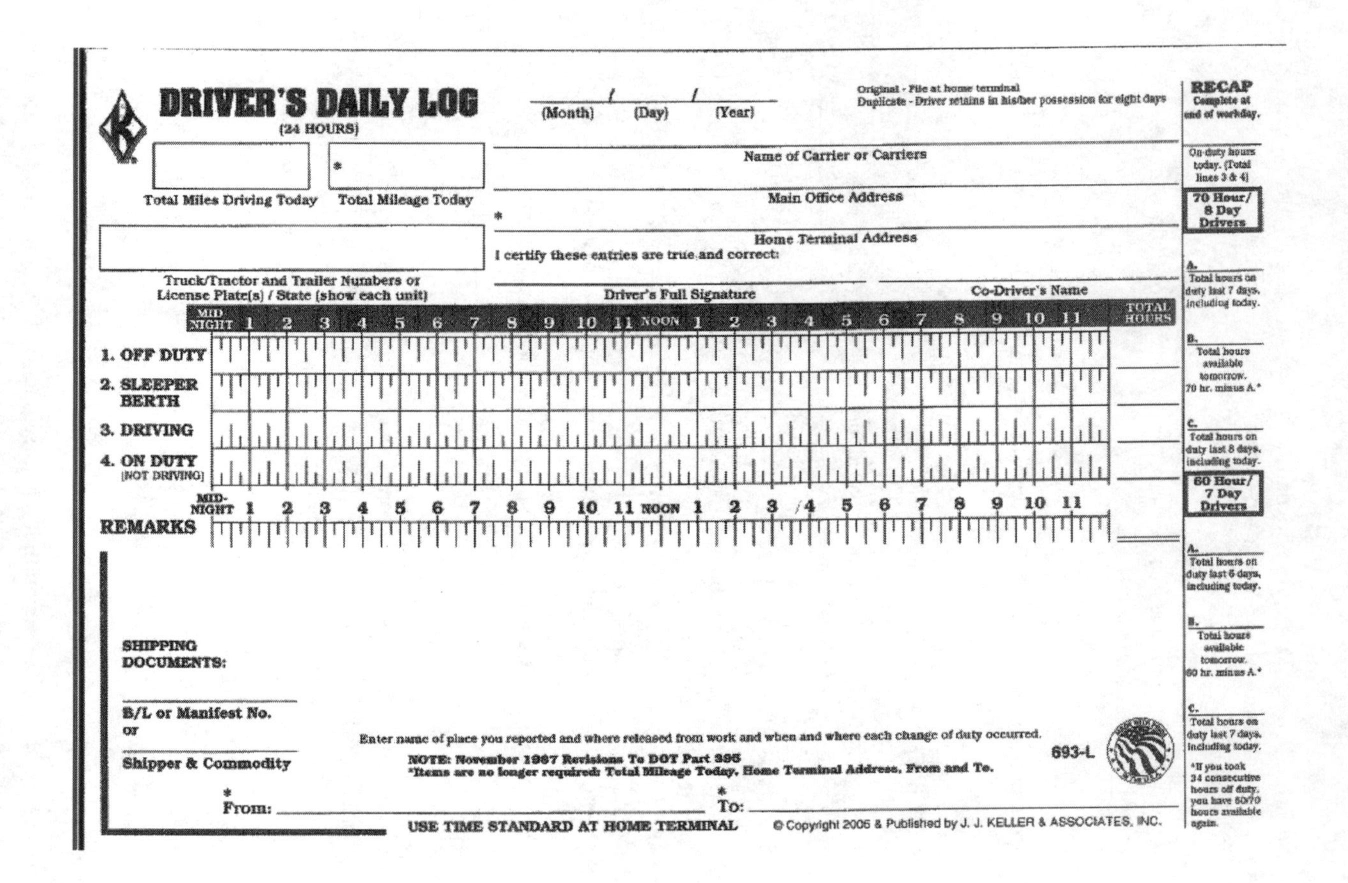

DRIVER'S DAILY LOG
(24 HOURS)

_____ / _____ / _____
(Month) (Day) (Year)

Original - File at home terminal
Duplicate - Driver retains in his/her possession for eight days

Total Miles Driving Today | * Total Mileage Today

Name of Carrier or Carriers

Main Office Address

* Home Terminal Address

Truck/Tractor and Trailer Numbers or License Plate(s) / State (show each unit)

I certify these entries are true and correct:

Driver's Full Signature | Co-Driver's Name

MID-NIGHT 1 2 3 4 5 6 7 8 9 10 11 NOON 1 2 3 4 5 6 7 8 9 10 11 TOTAL HOURS

1. OFF DUTY
2. SLEEPER BERTH
3. DRIVING
4. ON DUTY (NOT DRIVING)

MID-NIGHT 1 2 3 4 5 6 7 8 9 10 11 NOON 1 2 3 4 5 6 7 8 9 10 11

REMARKS

SHIPPING DOCUMENTS:

B/L or Manifest No.
or
Shipper & Commodity

Enter name of place you reported and where released from work and when and where each change of duty occurred.

NOTE: November 1967 Revisions To DOT Part 395
*Items are no longer required: Total Mileage Today, Home Terminal Address, From and To.

693-L

* From: _____ * To: _____

USE TIME STANDARD AT HOME TERMINAL

RECAP
Complete at end of workday.

On-duty hours today. (Total lines 3 & 4)

70 Hour/ 8 Day Drivers

A. Total hours on duty last 7 days, including today.

B. Total hours available tomorrow. 70 hr. minus A.*

C. Total hours on duty last 8 days, including today.

60 Hour/ 7 Day Drivers

A. Total hours on duty last 6 days, including today.

B. Total hours available tomorrow. 60 hr. minus A.*

C. Total hours on duty last 7 days, including today.

*If you took 34 consecutive hours off duty, you have 60/70 hours available again.

FIGURE 7.1 Driver's daily log form.

PREVENT ACCIDENTS BEFORE THEY OCCUR

The loss of human life is the most serious implication of an accident from a motor vehicle in a company's motor fleet. Injury and property loss are also motivating factors for ensuring an effective safety program. According to the *Motor Fleet Safety Manual*, the supervisor must be able to identify substandard performance that leads to accidents by doing the following:[2]

- Personally observing the driver's performance
- Checking training reports, road patrol reports, arrest records, and employee comments
- Immediately reviewing complaints from other drivers and pedestrians
- Being alert to changes in the driver's personality

MOTIVATING DRIVERS TO GET DESIRED RESULTS

Many companies rely heavily on two common criteria for employment: the maximum number of permissible moving violations and the number of accidents in the previous two years.

The best method of screening is to disclose these criteria up front and indicate that the company will obtain a *motor vehicle report* (MVR). This report gives an accurate and updated history of an individual driver. However, one pitfall that companies encounter (according to an article in *Professional Safety*) is that they fail to check the report again after hiring.[3] The article recommends making an MVR check part of your written program policy and checking driving records periodically, at least annually. You may also require employees to report all driving incidents when they occur. In fact, the Federal Motor Carrier Safety Regulations Part 381.23 states that "each motor carrier shall make the following investigations and inquiries with respect to each driver it employs…

1. An inquiry to each State where the driver held or holds a motor vehicle operator's license or permit during the preceding 3 years to obtain that driver's motor vehicle record.
2. An investigation of the driver's safety performance history with Department of Transportation regulated employers during the preceding three years."

Another practice that should be utilized is to categorize drivers based on their level of risk. It is an unavoidable fact that each fleet is made up of a diverse group of drivers, with varying risk levels. To determine your greatest risk, you must develop a risk profile for each driver. On average, 20 percent of the drivers are responsible for 80 percent of the accidents. By honing in on these high-risk drivers, a company can improve accident records and maximize the fleet safety return on investment.

SYMPTOMS OF ACCIDENTS THE SUPERVISOR SHOULD KNOW

It may appear wasteful to focus on drivers who have no track record of problems; however, statistics prove otherwise. An accident is most likely to occur during the first 18 months of a driver's tenure with a company. It also has been reported that new hires are responsible for 30 to 40 percent of fleet accidents.[3] Reasons why there is such a high incident rate for new employees include the following:

- New hires are busy learning about the company's products.
- New hires are trying to provide all services and policies required by the company.
- New hires are driving vehicles that differ from their personal vehicles.
- New hires have a tendency to overschedule and rush.

Particular attention should be paid to new hires with little driving experience or who have come from driving schools. The following points show a need for additional driver training.

- New hires may not receive uniform training.
- They may lack experience identifying inspection items that could lead to out-of-service violations.
- They may require more time in training.
- They can make simple mistakes that lead to expensive costs.

Driving Abuses That Lead to Undue Wear on Vehicles

A commercial vehicle with a gross vehicle weight ratio (GVWR) of 10,001 lbs. or more must be inspected, maintained, and repaired periodically. A more thorough explanation can be found in the Federal Motor Carrier Safety Administration's Regulatory Guidance Part 396—Inspection, Repair, and Maintenance.

The Federal Motor Carrier Safety Administration was established within the Department of Transportation on January 1, 2000, pursuant to the Motor Carrier Safety Improvement Act of 1999 (Public Law No. 106-159, 113 Stat. 1748 [December 9, 1999]). It was formerly a part of the Federal Highway Administration. The Federal Motor Carrier Safety Administration's primary mission is to prevent commercial motor vehicle–related fatalities and injuries.4 Administration activities contribute to ensuring safety in motor carrier operations by enforcing safety regulations, targeting high-risk carriers and commercial motor vehicle drivers, improving safety information systems and commercial motor vehicle technologies, strengthening commercial motor vehicle equipment and operating standards, and increasing safety awareness.

To accomplish these activities, the administration works with federal, state, and local enforcement agencies; the motor carrier industry; and labor safety interest groups, among others.

Another source of information can be obtained through the Commercial Vehicle Safety Alliance. To achieve a rating of "satisfactory" in the vehicle category, carriers must have no critical or acute violations, and their out-of-service rate must be less than 34%.

Driver Inspection

Daily inspections are a necessity when a company implements a fleet safety program. Responsibility is placed on the driver, mechanics, and fleet manager. The specific parts to be inspected are parking brakes, steering mechanisms, lights, tires, horns, windshield wipers, mirrors, coupling devices, wheels and rims, and emergency equipment. The constant abuse from everyday driving as well as seasonal weather changes may increase the necessity of additional inspections.

Part 396(b) of the Code of Federal Regulations states that "The report shall identify the motor vehicle and list any defect or deficiency discovered by or reported to the driver which would affect safety of operation of the motor vehicle or result in its mechanical breakdown. If no defect or deficiency is discovered by or reported to the driver, the report shall so indicate. In all instances, the driver shall sign the vehicle inspection report. On two-driver operations, only one driver needs to sign the report, provided both drivers agree with the report of the defects or deficiencies. If the driver operates more than one vehicle during the day, a report shall be prepared for each vehicle operated."[4]

POLICY DEVELOPMENT

A supervisor's attitude toward safe driving will greatly affect the attitude and driving performance of those responsible to him or her.[5] This should be carried over into a clearly worded policy that spells out the company's objectives and describes:

- How the company intends to comply with all safety laws and ordinances
- Safety of employees, the public, and equipment operations
- Commitment to safety, which takes precedence over expediency

Holding supervisors as well as drivers accountable for safe performance is another means of ensuring fleet safety. To hold someone accountable, a company must know whether he or she is performing well, so it must measure that person's performance. Without measurement, accountability becomes an empty and meaningless concept.[6] To ensure that accountability is maintained, a company must keep open the lines of communication between management and drivers, and barriers should not be placed between employees and management.

SAFETY MEASUREMENT FOR FLEET MANAGERS

According to *Techniques of Safety Management*, "Measurement is more crucial at the supervisory level, and the measure (which is the motivator here) must do many more things than at the employee level."[6] This principle can be applied directly to the fleet supervisor. The book lists a few criteria for good measurement of the fleet performance:

- It should give swift and constant feedback.
- It should get the supervisor's attention.
- It should measure the presence of safety activity, not only its absence (as indicated by accidents).
- It should be sensitive enough to indicate when effort has slowed.
- It should provide measures of good or poor performance.
- It should be meaningful.

Once a system of measurement has been established, it is much easier to communicate the effectiveness of a program to employees and, just as important, to upper management. Providing meaningful data on injury rates, accident frequencies, and property damage is critical when funding decisions must be made.

STUDY QUESTIONS

1. What are two common criteria that companies use when selecting drivers for employment?
2. What are four reasons new hires have a higher incident rate than tenured drivers?
3. What are five criteria that a fleet manager must look at when developing a process to measure the program's success?
4. When you are selecting a driver during the pre-employment interview, what are some areas that indicate past driving performance?
5. (True or False) Daily vehicle inspections are not necessary when a company implements a fleet safety program.
6. What does CSA stand for and what seven criteria are measured in the BASIC scores?

REFERENCES

1. Cullen, D., Safety is no accident, *Fleet Owner* 93, no. 4, 49–53, April 1998.
2. Brodbeck, J.E., ed., *Motor Fleet Safety Manual*, 4th ed., National Safety Council, Itasca, Illinois, 1996.
3. Moser, P., Rewards of creating a fleet safety culture, *Professional Safety* 46, no. 8, 39–41, August 2001.
4. Code of Federal Regulations, Title 49, Part 394, US Department of Transportation.
5. Rickey, J., Jr., Fleet safety guidelines, *Beverage Industry* 88, no. 8, 59–61, August 1997.
6. Petersen, D., *Techniques of Safety Management,* American Society of Safety Engineers, Des Plaines, Illinois, 1998.
7. Newnam,S., Griffin, M.A., and Mason, C., Safety in work vehicles: A multilevel study linking safety values and individual predictors to work-related driving crashes, *Journal of Applied Psychology* 93, no. 3, 632–644, 2008.
8. Federal Motor Carrier Administration, Safety measurement system. In *CSA—Compliance, Safety, Accountability*. http://csa.fmcsa.dot.gov/about/basics.aspx.
9. http://www.fmcsa.dot.gov/rules-regulations/rules-regulations.htm.
10. http://www.dot.gov/.

8 Motor Fleet Inspection Program

EMPLOYER RESPONSIBILITIES

The Occupational Safety and Health Act of 1970 states that every business must keep its employees as safe as possible from physical hazards. Keeping a work environment free of hazards is a nearly insurmountable task. Doing so requires careful use of the management tools of organizing, planning, leading, evaluating, and controlling. Every task, workstation, tool, operation, and machine must be carefully examined for safety issues. Working toward this end takes input from every employee in the organization. Everyone has an equal responsibility to remain alert and up to date in hazard detection. Equally important is a total commitment to support the motor fleet safety program, and everyone in the organization must aid in this effort as well. Only a program that has the total commitment of all employees will be successful.

A successful employee working in the motor fleet safety program is one who is flexible enough to change when conditions warrant that change. An efficient monitoring program provides the basic information management needs to ensure the program's effectiveness.[1]

If management prefers to see results based on the immediate cause of the problem, it should be very easy to identify the changes that are in order. If management prefers to see findings in print, the employee should carry out this task as soon as possible. If this task involves an accident, a complete accident investigation form should be covered in detail by summarizing the findings to include information about what damage, injury, and losses could occur if a similar accident were to be repeated. This creates urgency and emphasizes prevention of future losses. It is up to the supervisor, who knows the workplace as well as anyone else, to point out the obvious. Management must tell it like it is at the time. When the problem has been uncovered and recommendations stated for the change, it is up to the members of the management team to see that the changes are implemented and communicated to the drivers and other employees.

ACCIDENT PREVENTION

A motor fleet employee safety program starts with accident prevention. Each employee and every member of management should keep the following in mind:[1]

- Every employee in the organization must assume personal responsibility for safety.
- Accidents can happen anywhere at any time.
- Accidents are caused by unsafe practices and unsafe conditions.
- Careful study by everyone is an effective way to prevent accidents.
- If unsafe practices and unsafe conditions are not eliminated, then they will continue to contribute to an unsuccessful program.

Developing an effective safety program involves four steps:[1]

1. Inspecting for hazards
2. Training all employees
3. Enforcing rules and following up
4. Maintaining interest and commitment to safety throughout the company

Initial Safety Inspection

Once management understands the importance of training everyone to identify hazards, the next step is to train employees in the best methods for spotting physical hazards.

One effective method is for every workstation area of the fleet to create detailed safety inspection checklists. Checklists are a good way to get people thinking about safety. Constant safety awareness is a crucial element of a successful program.

Techniques Used in Initial Safety Inspections

When you are conducting the initial inspection, it is important to follow these steps to keep the process free from snags and resistance:[1]

- Anticipate mechanical and physical hazards.
- Let the inspection follow the job process.
- Do not distract employees while inspecting.
- Teach safe procedures when necessary.
- To accurately detect hazards, remember to use your senses.

If these points are respected, then the safety program will have a solid foundation.

Periodic Safety Inspections

After a system-wide initial inspection, periodic safety inspections follow. Periodic inspections keep the safety program up to date and consistent with the company's safety philosophy.

TRAINING

Employee training is a critical area of the motor fleet safety program. It is extremely important to establish a comprehensive safety training program. The training program should focus on key safety issues that need attention. Constant measurements of safety performance should be taken, and training sessions should be conducted for the areas that need additional training.

Driver Training

A variety of classroom, behind-the-wheel, self-study, and custom-tailored training programs must be offered to help reduce the chance that drivers will become involved in accidents. A safety manager can do this through a combination of materials:

- Video and live presentations by experienced instructors
- Facilitation packages that include videos, workbooks, and other classroom materials
- Behind-the-wheel training that supplements classroom training sessions
- Videos developed by the National Safety Council and other distributors of motor fleet programs

Follow-Up and Enforcement

Another important aspect of an effective program is maintaining the program's success. One way to do this is through follow-ups and the enforcement of safety policies.

Follow-up involves identifying any obstruction to the removal of hazards. If a hazard is not carefully eliminated, it is bound to appear again to cause problems. Areas of concern must be addressed using a follow-up protocol.

Enforcement is another crucial area of the motor fleet employee safety program. Everyone in the organization must take the program seriously. One minor instance of insubordination can be the catalyst to destroy all the hard work it took to get the program moving toward its goals. Enforcement of policy statements must be visible to everyone to send the message that safety is as important as production.

Maintaining Interest in the Safety Program

To maintain interest in the safety program, the company must gain the total involvement of everyone in the organization. Responsibility for safety should be given to everyone, and all employees need to contribute. If someone does not have a part in the safety program, that individual will lose interest, and awareness of hazards will drop to an unacceptable level. Interest will fall to a level where incidents are imminent.

FLEET ENTERPRISE JOB HAZARDS

In this section we discuss hazards that are typical to motor fleet safety. All of these hazard checks should be included in a company's inspection checklist as well as its training program. They should be reinforced to the point where every employee can immediately recognize and eliminate hazards without delay. Areas in which to look for common hazards are:[1]

- Eyesight
- Welding equipment
- Health
- Power hand tools
- Lifting: auto, bus, truck
- Explosive vapors
- Burns: chemical, water, steam
- Vehicle movement
- Heavy parts: removal, replacement
- Shop, station, garage
- Mounting: tires and rims
- Terminals
- Machine tools
- Truck and bus operations
- Slippery and uneven floors

These are areas where the greatest typical hazards exist and are good places to start to build a successful safety program.

Metalworking Tools

Power transmission tools, such as pulleys, sheaves, belts, gears, and clutches, are all hazards because they are under pressure. If the pressure is released, the object becomes a potential airborne, whipping, or hammering hazard.

Only workers who are properly trained should be in the point-of-operation zone. Unauthorized workers who are in a point-of-operation zone are a hazard to themselves and to those who work in those areas.

Cleaning Fluids

Cleaning fluids should be clearly labeled, capped, and stored in a cool, well-ventilated place away from sparks and flames. Typical cleaning fluids are kerosene and various safety solvents.

STIMULATING INTEREST IN THE SAFETY PROGRAM

Here are some ways to stimulate interest in the employee safety program. Creativity is encouraged to keep workers focused on safety:[1]

- Management interest and example
- Safety meetings and rallies
- Safety awards
- Safety contests
- Posters
- First aid training
- Safety suggestion systems

Display of posters must be planned carefully in order for the posters to be effective. There is limited space to exhibit them, and the posters must be eye-catching and informative. Pictures are effective if they clearly explain the message that is being sent. Posters should focus on one particular safety issue and be displayed in a relevant work area. Topics for posters include slips and falls, personal protective equipment, and various potential hazards. If properly executed, the display of posters is an effective way to keep safety awareness at a high level.

GENERAL SAFETY RULES

General safety rules are the final part of the motor fleet employee safety training program. These are general topics that should be familiar to every employee and manager. Remember:

- Safety is everyone's responsibility.
- Everyone should know where the first-aid materials are located, and they should be kept stocked.
- All employees should be trained in the company's fire plan.
- Electrical equipment must be labeled, and lock-out/tag-out procedures should be enforced.
- Work attire should include steel-toe boots and fire-retardant clothing.
- Personal protective equipment (PPE) should be in good condition and readily available.
- The work environment should be clean and free of debris.
- Overloading should never be practiced. Overloading is placing more cargo on a truck than it is legally permitted to carry or designed and powered to carry.

If these general areas are covered, then attention can be focused on making the safety program a success. However, keep in mind that a complete motor fleet employee safety program starts with commitment and effort from everyone in the organization.

VEHICLE INSPECTION

Inspection, Repair, and Maintenance

Every motor carrier should inspect, repair, and maintain all motor vehicles subject to its control.[1,2]

Parts and accessories should be in safe and proper operating condition at all times. This includes frame and frame assemblies, suspension systems, axles and attaching parts, wheels and rims, and steering systems.[1]

At least every 90 days, push-out windows, emergency doors, and emergency-door-marking lights in buses should be inspected.[1,3]

For all vehicles owned or operated for 30 consecutive days or more, except for a private motor carrier for passengers (nonbusiness), the motor carrier should maintain the following records:[1]

- An identification of the vehicle, including company number if so marked, make, serial number, year, and tire size. If the motor vehicle is not owned by the motor carrier, the record should identify the name of the person furnishing the vehicle.
- A means to indicate the nature and due date of the various inspection and maintenance operations to be performed.
- A record of inspection, repairs, and maintenance indicating their date and nature.
- A record of tests conducted on push-out windows, emergency doors, and emergency-door-marking lights on buses.

Maintenance

Every vehicle in the motor fleet industry has its own particular maintenance requirements and schedules to follow. Because maintenance is a very labor-intrinsic part of motor fleet operations, the technicians should be trained to operate the diagnostic equipment in order to maintain a safe, reliable, and cost-effective maintenance program. The objective of the maintenance shop is to reduce costs, improve the utilization of the vehicles, and increase productivity and profitability for the motor fleet operation.

Record Retention

The records should be retained where the vehicle is either housed or maintained for a period of 1 year. Records must be retained for 6 months after the motor vehicle leaves the motor carrier's control.[1]

Record-Keeping Requirements

Qualified inspectors will prepare a report that:

- Identifies the individual performing the inspection
- Identifies the motor carrier operating the vehicle
- Identifies the date and vehicle inspected
- Identifies vehicle components inspected and describes results of the inspection
- Identifies components that meet minimum standards
- Certifies accuracy and completeness of inspection

The company shall:

- Maintain the original or a copy of the inspection where the vehicle is housed or maintained
- Keep the inspection report for 14 months
- Make available the original or a copy of the inspection report on demand of an authorized federal, state, or local official[1]

Lubrication

Every motor carrier should ensure that each motor vehicle subject to its control is properly lubricated and free of oil and grease leaks.[1]

Inspection of Motor Vehicles in Operations

Only authorized personnel can perform inspections. Every special agent of the Federal Highway Administration (FHWA) is authorized to perform inspections of motor carrier vehicles in operation.[3]

The driver equipment compliance check should be used to record results of motor vehicle inspections conducted by authorized FHWA personnel. Motor vehicles declared out of service should not be used until fixed. Authorized personnel will declare and mark "Out of Service" on any motor vehicle that by reason of its mechanical condition or loading would likely cause an accident or a breakdown. An out-of-service vehicle sticker should be used to mark such vehicles. No motor vehicle carrier will require or permit any person to operate, nor will any person operate any motor vehicle declared and marked out of service, until all required repairs have been satisfactorily completed. The term "operate" as used in this section should include towing the vehicle; however, vehicles marked out of service may be towed away by means of a crane or hoist. A vehicle combination consisting of an emergency towing vehicle and an out-of-service vehicle should not be operated unless such a combination meets the performance requirements of the driver equipment compliance check.

Employees are not permitted to remove the out-of-service sticker from their motor vehicles. Only authorized mechanics may remove the sticker after all repairs required by it are completed.

Motor Vehicle Inspection Results

When an inspection has been completed, the following steps should be carried out:[4]

- The driver of any motor vehicle receiving an inspection report should deliver it to the motor carrier operating the vehicle upon his or her arrival at the next terminal or facility. If the driver is not scheduled to arrive at a terminal or facility of the motor carrier operating the vehicle within 24 hours, the driver should immediately fax, e-mail, or mail the report to the motor carrier terminal.
- Motor carriers will examine the report. Violations or defects noted need to be corrected. Red tags, methods of giving priority to a truck in need of repairs, may be indicated.
- Within 15 days following the date of the inspection, the motor carriers will certify that all violations noted have been corrected by completing the "Signature of Carrier Official, Title, and Date Signed" portions of the form, and return the completed roadside inspection form to the issuing agency at the address indicated on the form. They must also retain a copy for their records.

BENEFITS GAINED BY STANDARDIZING FLEET VEHICLES

Standardizing fleet vehicles is important to ensuring a well-operated program. Fleet programs are usually safer if all vehicles are similar or even the same. Proficiency is gained when an individual—for example, a mechanic—works on the same kind of vehicle every day. Parts can be purchased in bulk, and the extra parts can be shelved for later use. Using the same types of vehicles has other benefits: Drivers will be familiar with all of the vehicles and will be able to drive any vehicle at any given time. A driver will feel comfortable switching to another vehicle if his or her vehicle is in the shop.

BUILT-IN VEHICLE COMPONENTS DIRECTLY RELATED TO SAFETY

Commercial motor vehicles must be equipped with specific standard equipment. Optional equipment or accessories are permitted only if these items do not adversely affect the operational safety of the vehicle. Essential items include the following:

- Lights
- Lamp mounting
- Stop lamps
- Brakes
- Brake tubing and hose
- Brake warning device
- Windshield
- Fuel system
- Coupling device
- Cargo security

Lights

Part 393 of the Code of Federal Regulations (CFR) specifies the required position, color, type of lamps, and reflectors for commercial motor fleet vehicles.[2] All lamps and reflectors for commercial motor vehicles manufactured after March 7, 1989, must meet the requirements of Federal Motor Vehicle Safety Standard (FMVSS) no. 108 (49 CFR 571.108) in effect on the date of manufacture. Certain trailers manufactured after December 1, 1993, must have retroreflective sheeting or additional reflectors to make them more visible to other motorists under conditions of reduced visibility. Lamps must light on inspection and when required during regular operation of the vehicle. Permanently secured lamps are required at all times except when temporary lamps are in use. Some examples are a drive-away or tow-away operation or the mounting on projection loads (temporary lamps must be securely attached). When service brakes are applied, all stop lamps on commercial vehicles must be activated.

Brakes

All commercial vehicle systems must be equipped with the following brake systems:

- 49 CFR 393.41 requires a service brake system that specifies braking and holding performance.
- 49 CFR 393.41 requires a parking brake system that meets the requirements of parking brake activation and a method of holding the brakes in the applied position.
- 49 CFR 393.52 requires an emergency brake system that consists of either (1) emergency features of the service brake system or (2) a system separate from the service brake system.
- Speed governors are required. These devices control the maximum rate of speed that a vehicle may be driven by regulating the performance of the engine.

Brakes acting on all wheels must be equipped on all commercial motor vehicles, with the following exceptions:[2]

- Any truck manufactured before July 25, 1980, with three or more axles is not required to have steering axle brakes.
- During drive-away or tow-away operations, vehicles being towed are not required to have brakes on all wheels.
- Any full trailer, semitrailer, or pole trailer weighing 3,000 lbs. or less, that does not surpass 40 percent of the towing weight of the vehicle, is not required to have brakes on all wheels.

When towing a trailer equipped with brakes, the vehicle must maintain the operation of the brakes on the towing vehicle in the event that the trailer breaks away from the towing vehicle.

Tubing and hoses are critical to the safety of your motor vehicle. The tubing and hoses must be properly installed to provide safe performance. Remember that when tubing is near the exhaust systems the high temperature can harm the hoses. Be sure to consider the proximity to the exhaust system when installing tubing and hoses.

Motor vehicles are installed with warning signals that activate when the brakes fail. The signal covers hydraulic brake systems, air brake systems, and vacuum brake systems.

Windshield Condition

The windshield on a motor vehicle must be free of cracks and discoloration.

Fuel Systems

Any part of the fuel system should not extend beyond the widest part of the vehicle.

Equipment Required by State and Interstate Commerce Commission Regulations

Emergency equipment is required for the safety of drivers. Trucks must carry emergency equipment that is regulated by state and federal laws. Fire extinguishers (a five-pound dry chemical unit; petroleum trucks carry a fifteen-pound type), flags, reflectors, port torches, portable electric red flasher warning devices, fuses, spare bulbs, and emergency reflective triangles are some of the required emergency equipment. Hazardous material trucks must be equipped with a large type of fire extinguisher. Truck drivers with a higher risk (e.g., chemical hazards) in general carry protective clothing, gas masks, cleanup materials, and a breathing apparatus. Hard hats, rubber gloves, rubber boots, and face masks are recommended. If your fleet is running in cold temperatures or in a remote area, it is also recommended that your drivers have first-aid kits, blankets, survival suits, and emergency food and water.

CONTRIBUTIONS A DRIVER CAN MAKE TO AN EFFECTIVE MAINTENANCE PROGRAM

It is important that every carrier prepare a posttrip inspection report at the end of each day. Every driver must write a report for each vehicle driven. This report must address the following items:

- Service brakes (including trailer brake connections)
- Parking (hand) brake
- Steering mechanism
- Lighting devices and reflectors
- Tires
- Horn
- Windshield wipers
- Rearview mirrors
- Coupling devices

- Wheels and rims
- Emergency equipment

The report must list any condition that the driver either found or had reported to him or her that would affect safety of operation or cause a breakdown. If no defect or deficiency was identified, the report should state that. The driver must sign the report in all cases. Before the driver can dispatch the vehicle, he or she must ensure that if the report shows defects or deficiencies, repairs have been made. Some deficiencies may not need immediate attention; however, the report must state that in writing. The carriers must keep the original posttrip inspection report and the certificate of repairs for at least 3 months after the date of preparation.

The driver must be satisfied that the motor vehicle is in safe operating condition before starting out on the road. If the last vehicle report notes any deficiencies, the driver must review and sign to acknowledge that necessary repairs have been completed.

ADVANTAGES GAINED BY ASSIGNING A VEHICLE TO AN INDIVIDUAL

There are many advantages to assigning a particular vehicle to an employee. When you lease a vehicle to an individual, that person is responsible for a specific vehicle, which puts pressure on him or her to keep the vehicle clean and in good condition. Knowing that you have to be in the same vehicle every day motivates you to keep it well maintained and reduces abuse of the vehicle. It is convenient from a record-keeping point of view to keep track of the vehicles and accidents or other incidents that occur. For this reason, assigning vehicles to employees will decrease operating costs.

CHARACTERISTICS OF FLEET MANAGEMENT PROGRAMS

Spur of the moment changes that occur on the job can often lead to employee disruption. High turnover rates resulting from drivers becoming unhappy with procedure changes as well as lack of experience with the current program indirectly increase operating costs due to mishandling of equipment. Gradual changes to employee policies have been shown to reduce these complications.

STUDY QUESTIONS

1. (True or False) Mechanics who work on the same vehicles tend to develop expert knowledge and can save the fleet money and valuable time.
2. List four vehicle components that are directly related to safety.
3. Petroleum trucks have to carry fire extinguishers that weigh how many pounds?
4. What is the definition of overloading?
5. (True or False) When assigning a vehicle to an individual, you will lower operating costs and also lower vehicle abuse to the vehicles.
6. What is the definition of a speed governor?

7. List three required pieces of equipment regulated under the state and Interstate Commerce Commission regulations for motor fleet vehicles.
8. What are the general safety rules discussed in this chapter that contribute to a safety training program?
9. List the four major steps of establishing a consistent motor fleet employee safety program.
10. What are five typical hazards that every employee should be able to instantly spot and eliminate?
11. During initial safety inspections, what are some key points to remember?

REFERENCES

1. Brodbeck, J.E., ed., *Motor Fleet Safety Manual*, 4th ed., National Safety Council, Itasca, IL, 1996.
2. http://www.cvsa.org/Inspections/inspection_Procedures/inspection_ procedures.html.
3. Code of Federal Regulations, Title 49, Part 396, U.S. Department of Labor, Washington, D.C.
4. *Federal Motor Carriers Safety Regulations* (Handbook). U.S. Department of Transportation, Parts 40–399. Neenah, WI: J.J. Keller & Associates, 2004.

9 Organizing Motor Fleet Accident Data

ACCIDENT INVESTIGATION

Why should the organization of fleet and motor vehicle accident data be an essential and necessary part of your motor transportation safety program? The reason is simple: accident prevention. Each year thousands of drivers are killed or injured in motor fleet crashes that could have been prevented in one way or another. These accidents affect not only the drivers but everyone else as well. Whether an individual is driving a company car on a business trip, driving a truck that is part of a fleet, or sending a child to school on a school bus, everyone is involved in motor fleet safety.

RESPONSIBILITIES OF THE FLEET SAFETY DIRECTOR

Most accidents have more than one cause. In accident prevention, the job of the fleet safety director is to determine all of the factors that lead to accidents and then attempt to control or eliminate these factors by planning ahead.[1] This planning for accident prevention is one of the many responsibilities of the fleet safety director, who will find it beneficial to incorporate accident data into this planning process.

The fleet safety director has two main responsibilities with regard to motor fleet safety. The first is identifying causes of accidents. If safety directors know what factors cause an accident, then they are halfway home; eliminating those factors is the next step. In that way, the chances of the same type of accident occurring again are greatly reduced. The second major responsibility of the fleet safety director is to recommend ways to remove the factors that cause accidents and to guard against accidents by protecting the employees. Drivers can be told that a simple thing to remember is "While behind the wheel, you are a driver. All attention needs to be on that aspect of your job when conducting company business. This is a time to drive with full concentration."

WHERE TO LEARN ABOUT CAUSES OF ACCIDENTS

PUBLISHED MOTOR VEHICLE ACCIDENT STATISTICS

Motor vehicle accident statistics have been collected and published by various organizations and are made available to motor fleet safety directors. One of the best sources available on the market is *Injury Facts* (formerly titled *Accident Facts*).[2] The National Safety Council publishes this book, and it contains a tremendous amount of useful statistics and information. It even breaks down the incidents by type of vehicle, work zones, states, and so forth.

The enterprise that employs a fleet safety manager might be a regulated motor carrier or a company that has a fleet of service-type or passenger vehicles. Based on this information, there are a number of resources that include the facts about collision causation. The following are such resources:

Bureau of Labor Statistics
Federal Motor Carrier Safety Administration
National Safety Council
National Highway Traffic Safety Administration
Occupational Safety & Health Administration
National and local trade associations
National School Bus Association
Insurance carriers

The organizations mentioned above have been of special interest to motor fleet safety directors for gathering background statistics and data to assist them when implementing training programs for fleet drivers.

SPECIAL STUDIES RELEASED BY INSURANCE COMPANIES

Studies that are released by insurance companies can be very helpful and informative. Insurance companies spend an astonishing amount of time and money investigating hazards and accidents. They do this to help eliminate accidents and thus save themselves money. The fleet safety director can also use these studies to help reduce accidents and help his or her company save money and possibly lives.

NATIONAL SAFETY COUNCIL

National Safety Council statisticians have created a comprehensive system for tracking and compiling injury and illness data, including annual publication of the aforementioned *Injury Facts*, an authoritative compendium of safety and health statistics. Council researchers also produce the *Journal of Safety Research*, an international, interdisciplinary scientific quarterly that contains research articles written by experts in all fields.[2] The council's Environmental Health Center, based in Washington, D.C., is a leading provider of credible and timely information and community-based programs on environmental and public health issues.

Other Sources

Other sources where a fleet safety director can find facts about causes of accidents are the Federal Highway Administration, the U.S. Bureau Labor of Statistics, summaries and reports of the Interstate Commerce Commission, and the National Institute for Occupational Safety and Health (NIOSH).

FEDERAL MOTOR CARRIER SAFETY REGULATIONS

According to 49 CFR 390.15, as noted in the *Federal Motor Carrier Safety Regulations Handbook*, motor carriers should keep all records of information pertaining to all accidents. These records must be maintained for a period of 1 year after an accident occurs and must be made readily available to any authorized representative or any special agent of the Federal Highway Administration.[3]

Information Needed

The following is taken directly from the *Federal Motor Carrier Safety Regulations Handbook*,[3] Title 49, Transportation Part 390.15—Investigations and Special Studies:

> (b) Motor carriers shall maintain for a period of one year after an accident occurs, an accident register containing at least the following information:
>
> (b)(1) A list of accidents containing for each accident:
>
> (b)(1)(i) Date of accident
>
> (b)(1)(ii) City or town in which or most near where the accident occurred and the state in which the accident occurred
>
> (b)(1)(iii) Driver name
>
> (b)(1)(iv) Number of injuries
>
> (b)(1)(v) Number of fatalities
>
> (b)(1)(vi) Whether hazardous materials, other than fuel spilled from the fuel tanks of motor vehicles involved in the accident, were released
>
> (b)(2) Copies of all accident reports required by state or other governmental entities or insurers

ACCIDENT RECORD SYSTEMS

All good accident record systems are organized into three distinct tasks: gathering the data, analyzing the data, and applying the data to develop countermeasures.

The first task, gathering the data, involves developing a carefully planned system for gathering all important data. This step is necessary for several reasons:

- To eliminate causes of accidents
- To avoid court action
- To improve labor relations
- To place responsibility where it rightfully belongs

The next task, analyzing the data, requires breaking down every accident into its component parts and tallying the information so that it can be viewed objectively.

The third step consists of applying the data to develop countermeasures. After we determine what causes were responsible, we must try to eliminate or reduce these factors in order to prevent the accident from happening again.

ACCIDENT INVESTIGATORS

Every company needs to have an accident investigator who can arrive at the scene of an accident quickly while keeping in mind the company's best interests. The accident investigator needs to be a take-charge person who can remain calm and collected. He or she might even have to direct traffic until other authorities arrive at the scene of the accident. It is important that the accident investigator be properly trained in the specifics of conducting a detailed investigation, which is the investigator's main responsibility. Other responsibilities are to prevent further damage at the scene, protect company property, protect the injured until medical aid arrives, and prevent excess costs.

An accident investigator's job is completed after he or she has determined exactly how the accident took place, identifies the factors that led to the accident, gathered all information and physical evidence, and completed all accident analysis records.

INFORMATION GATHERED IN ACCIDENT ANALYSIS RECORDS

To provide a complete breakdown of an accident and to pursue a step-by-step accident investigation, the investigator needs to gather the following information:[4]

- Time of the accident
- Location of the accident
- Description of what was involved in the accident
- Classification of accident
- Drivers' information
- Types of vehicles
- List of injured victims
- Witness information
- Movement of vehicles, pedestrians, or passengers
- Conditions
- Contributing factors
- Accident diagram
- Drivers' accounts of accident
- Drivers' suggestions
- Photographs

CLASSIFICATION OF MOTOR VEHICLE TRAFFIC ACCIDENTS

The purpose of ANSI-16, *Manual on Classification of Motor Vehicle Traffic Accidents*, from the American National Standards Institute (ANSI), is to provide a common language for those collecting and using traffic crash data.[4] The following is taken directly from the manual:

2.6 Road Vehicle Accident Types

2.6.1 Overturning accident: An overturning accident is a road vehicle accident in which the first harmful event is the overturning of a road vehicle.

2.6.2 Collision accident: A collision accident is a road vehicle accident other than an overturning accident in which the first harmful event is a collision of a road vehicle in transport with another road vehicle, other property or pedestrians.

2.6.3 Non-collision accident: A non-collision accident is any road vehicle accident other than a collision accident.

- Inclusions:
- Overturning accident
- Jackknife accident (See 2.6.4.)
- Accidental poisoning from carbon monoxide generated by a road vehicle in transport
- Breakage of any part of a road vehicle in transport, resulting in injury or in further property damage

2.6.4 Jackknife accident: A jackknife accident is a non-collision accident in which the first harmful event results from unintended contact between any two units of a multiunit road vehicle such as a truck combination.

2.6.5 Collision involving pedestrian: A collision involving pedestrian is a collision accident in which the first harmful event is the collision of a pedestrian and a road vehicle in transport.

2.6.6 Collision involving motor vehicle in transport: A collision involving motor vehicle in transport is an accident that is both a motor vehicle accident and a collision accident in which the first harmful event is the collision of two or more motor vehicles in transport.

2.6.7 Collision involving other road vehicle in transport: A collision involving other road vehicle in transport is an accident that is both an other road vehicle accident and a collision accident in which the first harmful event is the collision of two or more other road vehicles in transport.

2.6.8 Collision involving parked motor vehicle: A collision involving parked motor vehicle is a collision accident in which the first harmful event is the striking of a motor vehicle not in transport by a road vehicle in transport.

2.6.9 Collision involving railway vehicle: A collision involving railway vehicle is a collision accident in which the first harmful event is the collision of a road vehicle in transport and a railway vehicle.[6,7]

2.6.10 Collision involving pedal cycle: A collision involving pedal cycle is an accident that is both a motor vehicle accident and a collision accident in which the first harmful event is the collision of a pedal cycle in transport and a motor vehicle in transport.

2.6.11 Collision involving animal: A collision involving animal is a collision accident in which the first harmful event is the collision of an animal, other than an animal powering an other road vehicle, and a road vehicle in transport.

2.6.12 Collision involving fixed object: A collision involving fixed object is a collision accident in which the first harmful event is the striking of a fixed object by a road vehicle in transport. Fixed objects include such objects as guardrail, bridge railing or abutments, construction barricades, impact attenuators, trees, embedded rocks, utility poles, ditches, steep earth or rock slopes, culverts, fences and buildings.

2.6.13 Collision involving other object: A collision involving other object is any collision accident other than a (1) collision involving pedestrian, (2) collision involving motor vehicle in transport, (3) collision involving other road vehicle in transport, (4) collision involving parked motor vehicle, (5) collision involving railway vehicle, (6) collision involving pedal cycle, (7) collision involving animal, or (8) collision involving fixed object.

STUDY QUESTIONS

1. What is the main reason for gathering accident data?
2. What step is crucial in determining the factors leading to an accident?
3. All good accident record systems are organized into three distinct tasks. What are these three tasks?
4. A carefully planned system for gathering all important data is necessary for four reasons. List those reasons.
5. What information should be gathered in the accident analysis records?

REFERENCES

1. Brodbeck, J.E., ed., *Motor Fleet Safety Manual*, 4th ed., National Safety Council, Itasca, Illinois, 1996.

2. *Injury Facts*, National Safety Council, Itasca, Illinois, 2011.
3. *Federal Motor Carriers Safety Regulations Handbook*, J.J. Keller & Associates, Neenah, Wisconsin, 2001.
4. http://www.ansi.gov.
5. Kobelo, D. and Moses, R., *Influence of Truck Lane Restriction on Crash Occurrence: The Florida Case Study*, Department of Civil Engineering, FAMU-FSU College of Engineering, Transportation Research Board 87th annual meeting Compendium of Papers CD-ROM 2008, Washington, D.C.
6. Federal Railroad Administration, 2008, http://safetydata.fra.dot.gov/officeofsafety/.
7. Saccomanno, F.F., Park, P.Y., and Fu ,L., Estimating countermeasure effects for reducing collisions at highway-railway grade crossings, *Accident Analysis & Prevention* 39, 406–416, 2008.

10 Job Safety Analysis

IDENTIFYING RISK AND JOB TASK

Occupational injuries and fatalities occur every day in the workplace. These injuries often happen because employees are not trained in the proper job procedure. One way to prevent workplace injuries is to establish proper job procedures and train all employees in safer and more efficient work methods.[1]

A job safety analysis (JSA) is a simple method for identifying risks associated with a job task. By identifying hazards, a company can reduce risks and develop appropriate safety procedures. This is an essential process in reducing occupationally related injuries.[2]

DEFINITION

A JSA is a method that can be used to identify, analyze, and record:[3]

1. The steps involved in performing a specific job
2. The existing or potential safety and health hazards associated with each step
3. The recommended action(s) or procedure(s) that will eliminate or reduce these hazards and the risk of a workplace injury or illness

A JSA is a program that examines significant activities associated with a particular job. When the job is examined, the objective is to use experienced worker input to identify actual or potential health or safety hazards. Once these conditions are identified, the company can develop and document recommendations to serve as an operational tool alerting workers to health and safety hazards. The JSA is a program that can be easily adapted to suit any sector. Some sectors are:[4]

- Industry
- Health care
- Public service
- Child care
- Transportation

CONDUCTING A JOB SAFETY ANALYSIS

Basic Steps in a Job Safety Analysis

In a typical JSA, the left-hand column shows the basic steps of the job, listed in the order in which they occur. The middle column describes hazards or potential accidents associated with each job step. The right-hand column lists the safe procedures that should be followed to safeguard against identified hazards and to prevent potential accidents.[5]

A JSA consists of four steps:

1. Select the job to be analyzed.
2. Identify the hazards and potential accidents.
3. Break the job down into successive steps or activities.
4. Develop ways to eliminate the hazards and prevent potential accidents.

Selecting the Job

Jobs with the worst accident history have priority and should be analyzed first. In selecting jobs to be analyzed and the order of analysis, supervisors should be guided by the following factors:[1]

1. *Frequency of accidents.* A job that has repeatedly caused accidents is a candidate for JSA. The greater the number of accidents associated with the job, the greater its JSA priority.
2. *Rate of disabling injuries.* Every job that has disabling injuries should be given a JSA.
3. *Severity potential.* Some jobs may not have a history of accidents but may have the potential for a severe injury.
4. *New jobs.* A JSA of every new job should be made as soon as possible. Analysis should not be delayed until accidents or near misses occur.
5. *Near misses.* Jobs where near misses or close calls have occurred also should be given priority.

Extremely broad jobs (remediating a hazardous waste site) or extremely narrow jobs (setting a switch) are not suitable for the JSA process. Examples of jobs suitable for JSAs are drum sampling or operating a specific piece of equipment. A job may include a variety of specific tasks that should be included as part of the JSA.[6]

Breaking Down the Job

To do a job breakdown, select the right employee to observe. Choose an experienced, capable, and cooperative worker who is willing to share ideas. Explain the purpose and the benefits of the JSA to the employee.

Observe the employee perform the job and write down the basic steps. You may want to videotape the employee performing the job so that the tape can be used for review in the future. To determine the basic steps, ask, "What step starts the job?" Then, "What is the next basic step?" and so on. Completely describe each step. Any deviation from the regular procedure should be recorded, because it may be this irregular activity that leads to an accident.

Number the job steps consecutively in the first column of the JSA. Each step should tell *what* is done, not *how* it is done. The wording for each step should begin with an action word such as *insert*, *open*, or *weld*. The action is completed by naming the item to which the action applies; for example, *insert board* or *weld joint*. Be sure to include every step of the job from beginning to end.[1]

Identifying Potential Hazards

The next step in developing the JSA is the identification of all hazards involved with each step. Close observation and knowledge of the particular job is required if the JSA is to be effective in defining job hazards. To ensure that all hazards associated with a task (job step) are identified, observers should examine hazards produced by both the work environment and the specific activity being performed. Ask yourself the following questions about each step:[6]

- Is there a danger of striking against, being struck by, or otherwise making harmful contact with an object?
- Can the employee be caught in, by, or between objects?
- Is there potential for a slip or trip? Can the employee fall from one level to another or even on the same level?
- Can pushing, pulling, lifting, bending, or twisting cause strain?
- Is the environment hazardous to safety or health? Are there concentrations of toxic gas, vapor, mist, fumes, or dust? Are there potential exposures to heat, cold, noise, or ionizing radiation? Are there explosive or electrical hazards?

All of these questions can be incorporated into an inspection form that can be filled out at regular intervals. Even if the question may not apply at first, it may become relevant if there is a change from the standard operating procedures. Using a checklist is a good way to be sure nothing is overlooked. Employers should develop a checklist for each operation.[1]

The following hazards should be considered when completing a JSA:[2]

- Impact with a falling or flying object
- Penetration of sharp objects
- Caught in or between a stationary or moving object
- The danger of being struck by motor vehicles
- Fall from an elevated work platform, ladders, or stairs
- Excessive lifting, twisting, pushing, pulling, reaching, or bending
- Exposure to vibrating power tools, excessive noise, cold or heat, or harmful levels of gases, vapors, liquids, fumes, or dusts
- Repetitive motion
- Electrical hazards
- Light (optical)
- Radiation
- Heat (e.g., welding)
- Water (potential for drowning or fungal infections caused by wetness)
- Violence

Developing Solutions

The last step in a JSA is to develop a recommended safe job procedure to prevent accidents. Here are some solutions that should be considered:

1. Find a new way to do the job.
2. Change the physical conditions that create the hazards.
3. Change the work procedure.
4. Reduce the frequency of the job or task.

If a new way to do the job cannot be found, then try to change the physical conditions (tools, materials, equipment, layout, or location) of the job to eliminate the hazards.

When changing the work procedure is the best solution, find out what the employee can do during the job to eliminate hazards or prevent accidents. Employees may be able to suggest ways to improve safety on their work site.

A repair or service job may have to be repeated because a condition needs correction again and again. To reduce the need for such a repetitive job, find out what can be done to eliminate the cause of the condition that makes excessive repairs necessary.

Reducing frequency of a job contributes to safety only in that it limits the exposure. Every effort should still be made to eliminate hazards and to prevent accidents by changing physical conditions, revising job procedures, or both.

List recommended safe operating procedures on the form and also list required or recommended personal protective equipment (PPE) for each step of the job. Be specific; say exactly what needs to be done to correct the hazard. If the hazard is a serious one, it should be corrected immediately. The JSA should then be changed to reflect the new condition.[1]

MONITORING AND REVIEWING JSAS

In reality, making a workplace safer requires the commitment and cooperation of each employee. The development and implementation of the JSA requires integrated effort and shared responsibilities. Successful application of the process to ensure safe performance of assigned work will occur when line supervisors, health and safety personnel, and individual employees share responsibility, as defined in this section.

LINE SUPERVISORS

Line supervisors are responsible for:

- Ensuring that JSAs are identified and developed for appropriate jobs in accordance with this standard
- Coordinating the development and maintenance of JSAs with the appropriate health and safety personnel
- Incorporating existing hazard assessment information into the JSA process as appropriate
- Coordinating training on jobs covered by JSAs for all employees who perform such jobs and ensuring that documentation of all related training is maintained
- Coordinating review and revision of existing JSAs to ensure appropriateness, reflect procedural changes, and incorporate lessons learned from work experience, including accident investigations or critiques

EMPLOYEES

Employees are responsible for:[6]

- Assisting with development and validation of JSAs
- Following instructions provided by the JSA and associated training
- Notifying line management of new conditions that could affect the performance of a job, thus impacting the JSA

The completed JSA is reviewed by the supervisor, in consultation with assigned employee(s), before the planned work activity begins. The employees assigned to the task are trained in the new safe system of work. This confirms an understanding of the scope of the work, levels of employee experience and capability, and planned arrangements for the control of hazards. It also allows further employee identification and analysis of hazard potential and the opportunity to initiate additional control measures.

Management needs to directly observe the safe system continuously to monitor its adequacy and application. Upon completion of the job, managers should review and update the JSA for future reference.[7]

BENEFITS OF A JSA

Establishing proper job procedures is one of the benefits of conducting a JSA: carefully studying and recording each step of a job, identifying existing or potential job hazards (both safety and health), and determining the best way to reduce or eliminate these hazards.[1] The JSA process is perhaps one of the most effective ways of enabling an individual employee to participate and work as part of a team. In fact, employees are able to make important contributions and, through continuous work practice improvement, establish best-practice initiatives.[7] The benefits of performing a JSA are many, including:[1]

- The opportunity to provide individual training in safe and efficient work procedures
- The development of employee safety contacts
- Preparation for planned safety observations
- The opportunity to trust new employees on the job
- Pre-job instruction for irregular jobs
- Review of job procedures after accidents occur
- Examination of jobs for possible improvements in job methods
- Identification of safeguards that need to be in place
- Opportunity for supervisors to learn about the jobs they supervise
- Design of ergonomically correct work tasks
- The chance for employees to participate in workplace safety
- Reduced absenteeism
- Lowered workers' compensation costs
- Increased productivity
- Positive attitudes about safety

STUDY QUESTIONS

The following are short-answer questions:

1. What is the definition of a job safety analysis?
2. What are the four basic steps in a JSA?
3. What four factors should guide supervisors while selecting jobs to be analyzed?
4. List five hazards that should be considered when completing a JSA.
5. List five benefits of a JSA.

The following require a True or False response:

1. A JSA can be easily adapted to suit any sector, such as child care and transportation.
2. An inexperienced employee should be selected to perform a job breakdown.
3. One responsibility an employee has is to ensure that JSAs are identified and developed for appropriate jobs.

4. One responsibility a line supervisor has is to follow instructions provided by a JSA and associated training.
5. Reduced absenteeism and lowered workers' compensation costs are two benefits of performing a JSA.

REFERENCES

1. *Job Safety Analysis Identification of Hazards*, Occupational Safety & Health Bureau, Montana Department of Labor and Industry, Helena, Montana, http://erd.dli.state.mt.us/ safety health/brochures/jobsafetyanalysis.pdf.
2. *Workplace Safety: Job Safety Analysis*, SafeSpaces Inc., http://www.safespaces.com/ SS.Safety_htm/Safety_JobSafety Analysis.htm.
3. *Job Safety Analysis (JSA)*, Wisconsin Department of Administration, http://www.doa. state.wi.us/dsas/risk/risk/riskdocs/job safana.doc.
4. *Job Safety Analysis*, Preventive Action Safety Services Ltd., http://www.autobahn. mb.ca/~passltd/job%20safety%20 analysis.htm.
5. Brodbeck, J.E., ed., *Motor Fleet Safety Manual*, 4th ed., National Safety Council, Itasca, Illinois, 1996.
6. *Job Safety Analysis*, GJO Health and Safety Standards, http://www.doegjpo. com/public/manuals/gjo2/gjo2–2-8.pdf.
7. *Job Safety Analysis: Introduction*, Work Safe Western Australia, http://www.safetyline. wa.gov.au/institute/level1/course6/ lecture17/ 117_01.asp.

11 Safety Meetings for Motor Carrier Drivers

INCIDENT REDUCTION

Safety meetings for motor carrier drivers are designed to help reduce the number of incidents that occur in the workplace, although it may appear that safety meetings do not seem to be very effective.[1] The reason for this is many employees have negative feelings about safety meetings and do not attend. The goal of a safety manager in the motor carrier industry is to help all operators do a better job.[2] This includes the following:

- Stimulate and maintain interest in accident prevention.
- Develop attitudes sympathetic to the motor carrier safety program.
- Educate and train drivers and supervisory personnel in every factor that enters into safe commercial vehicle operation.

By finding ways to include these ideas, safety managers can improve the effectiveness of safety meetings and concentrate on the objectives that they are trying to accomplish with the group.

Developing objectives is an important aspect of conducting meetings. When setting individual and/or group objectives, these goals become a measuring stick of the effectiveness of what the safety manager is trying to accomplish. Each group meeting should have some basic goals that participants are attempting to reach. This list is universal regardless of subject:

- To improve the efficiency of all operations
- To establish a common bond of interest
- To instill in each driver a closer feeling of identity with the group

By seeking to incorporate these objectives, a safety manager strives to enhance employee morale. This may lead to an improved safety and work culture. The purpose is to leave employees with a sense of unity that results in improved performance and a potentially better life experience in the workplace.

TYPES OF MEETINGS

Once these objectives have been developed and incorporated into meeting materials, the safety manager should determine the type of meeting he or she will administer and decide which strong points are needed to get the message across to the drivers. Safety meetings can be categorized as one of the following:[3]

1. *Inspirational meeting*: Usually emphasis is placed on courtesy, goodwill, and loyalty to the group, or humanitarianism. The objective of these meetings is to exert a favorable safety influence by encouraging employees to develop a greater civic and moral responsibility.
2. *Celebration of accomplishments in accident prevention*: Celebrations of accomplishments in accident prevention meetings are usually held at the same time bonuses and awards are presented. The theme closely parallels that of inspirational meetings.
3. *Group instruction and training*: Group instruction and training meetings are held to train drivers in company policies and the factors that enter into the safe and efficient handling of business. Drivers are instructed in the key elements of safe, courteous, and efficient driving.

Along with determining the type of meeting that is necessary to help you get the desired employee response, you must consider other factors when planning a safety meeting. First, *scheduling* is important because if you schedule your meeting on a weekend or late at night, you are definitely going to have poor attendance. In addition, employees do not want to travel great distances to attend a meeting, so hold meetings at, or close to, the work site. When you are considerate to your employees, they are more willing to yield to your request.

SUBJECTS OF MEETINGS

The key to an effective meeting is presenting a subject that is exciting or so important that employees feel that it is in their best interest to pay attention. If a subject is really bland or dry, it may lead employees to skip the meeting, which is a contributing factor to program ineffectiveness. Some subjects that may be considered are:

- Thorough explanation of the public relations elements of the enterprise

- Traffic rules and regulations: explanation of the purpose and intent of regulations, discussion of standard signs and signals and their significance
- Instruction in driving practices, courtesy, and safety in general
- Mechanics of the vehicle: demonstration of care and maintenance by mechanics or specialists
- Explanation of physical problems: reaction time, stopping distance, skidding, passing distance, force of impact, etc.
- Accidents: past accidents, common situations, unusual accidents, seasonal hazards, what to do in case of accidents, report forms, frequency and severity of major types of accidents
- First aid and general health: importance, value of regular physicals, fatigue, illness, company policy concerning these areas
- Loading and unloading: how to lift, proper storage of cargo, handling of passengers
- Necessity for the proper use of special equipment such as flares, flags, chains, fog lights, spot lights, tarpaulins, air brake equipment, lighting equipment and connections, etc.
- Interstate Commerce Commission: state, city, or special regulations
- Open forum: questions to be asked and answered

These procedures and many more may find their way into your meeting. If there is a procedure you feel uncomfortable presenting because of your lack of knowledge on the subject, then do not hesitate to ask someone more knowledgeable to present the subject matter thoroughly and completely. Hiring a professional speaker is one option. Types of speakers include:

- Specialists and instructors: Instill confidence and convince listeners of the benefits of safety.
- Company officials: Display management's interest and support with inspirational addresses, future company plans, and a focus on the employees' welfare.
- Insurance carrier representative: Provides expert advice on the subject of accidents. Insurance companies offer this service.

These types of presenters may increase your chances of holding your audience's attention, which is a good idea for many reasons. Your employees are listening, and it increases the possibility of retention. The audience will remember what this meeting was about and the importance of safety. Other factors that play an important role in making your safety meetings a success are:

- Quiet environment
- Ample number of chairs
- Orderly arrangement of room
- Comfortable room temperature
- Adequate lighting
- Provisions for hat and coat placement
- Drinking water for speakers and others

Although these types of details seem insignificant, they help make the audience feel comfortable and important, along with putting the speaker at ease as well. When the speaker and the audience feel relaxed, the chances for a successful meeting increase significantly. Another idea is to praise those meeting participants who deserve it. You should praise people twice as much as you criticize them. Never let any good deed or action go unheralded in the group or organization. Also, you should try to give everyone a voice in the process. Nothing shows a group that you care for them more than listening to what they are trying to emphasize. So listen and consider what your workers have to offer during these important and informative meetings.

SUMMARY

Any organization, whether in manufacturing or transportation, should have safety meetings. When you start holding safety meetings, your employees will show their appreciation for your concerns. Although it is not possible for an organization to be entirely incident free, the only way you can make your employees more aware is through education and training. Learning how to conduct effective safety meetings is important to you as the fleet safety director.

STUDY QUESTIONS

1. Name the three types of meetings and give a brief description of each, including the purpose.
2. Name the three different types of speakers.
3. In a short paragraph, explain why the small things are important in the meeting room, such as the number of chairs, temperature of the room, and so forth.
4. Name two of the three reasons why objectives are important when conducting a safety meeting.
5. List two factors that play an important role in making safety meetings successful.

REFERENCES

1. Brodbeck, J.E., ed., *Motor Fleet Safety Manual*, 4th ed., National Safety Council, Itasca, Illinois, 1996.
2. *Fleet Risk Management and Driver Training Benefits*, Driving Services, http://www.drivingservices.com/benefits.htm.
3. Della-Giustina, D.E., *Developing a Safety and Health Program*, CRC Press, Boca Raton, Florida, 2000.

12 Motor Fleet Transportation Publicity

IMPROVING TRAFFIC SAFETY

An individual can influence traffic safety in many ways. Almost every person in the United States is either a driver or a passenger in a vehicle every day. This means that almost every person in the United States affects traffic safety each day. The choice to improve traffic safety lies with everyone on the road.

ROLES OF PROFESSIONALS AND PRIVATE CITIZENS

Traffic safety involves both professional motor carrier drivers and private motorists. Professional driver and fleet safety representatives need to understand the importance of safety on the road. Being a safety-minded driver, professionals should always act in a way that reflects this understanding. Taking the initiative to act in a safe manner on the highways is the responsibility of each person on the road, especially drivers who have been trained on the importance of safety on the highways.

It is the responsibility of all professional motor carriers driver as well as anyone who recognizes the importance of safety on the roadways to enforce safety policies, procedures, and regulations. Unsafe drivers endanger not only themselves but other drivers on the road as well.[4]

FLEET EXPECTATIONS

Any driver or fleet manager is required to know the specific process that is unique to his or her field. For example, someone dealing with the transportation of hazardous substances is expected to abide by the national standards. Management is responsible for providing additional training to employees working with hazardous materials (hazmat). Employees should be warned of hazards that are unique to their jobs. An example is the transportation of biochemicals, flammable substances, or materials that may act adversely in air or water. Knowing what the potential hazards are and how to handle them gives the professional driver a chance to survive if something goes wrong. (See Chapter 14 for more information on this topic.)

Being a Good Driver

When on the road, good drivers act in a way that they would want someone else to act. Driving in this manner will gain the respect and trust of fellow employees as well as the public. A good driver does the following:[1]

- Avoids accidents
- Performs pre- and posttrip inspections
- Avoids abrupt starts and stops
- Avoids schedule delays
- Does not irritate the public
- Performs the nondriving parts of the job
- Finds satisfaction in the job
- Gets along with others
- Follows traffic regulations and road markings
- Adapts to meet existing conditions

If you can find employees who fit this profile, they will become valuable assets. A good driver should be treated with respect and should not be taken for granted. Management should make an effort to keep and reward good drivers. This will ensure the success of the motor fleet organization.

All motor vehicle drivers should know that traffic safety efforts will affect their safety as well. Drivers should be up to date on any potential changes that will affect safety on the highways. After they have received this information, drivers should actively support the safety measures.

KEY ELEMENTS IN DRIVER TRAINING

Training gives a driver the knowledge and the skills to do his or her job properly, as well as an appreciation for the job's importance and the ability to do the job safely. The only way to ensure top driving skills is to train each new employee. Management should not assume that previous training is sufficient. Organizations realize many positive benefits by training drivers well.

Effective training means that a group will be more efficient and accidents will be eliminated, or at least reduced. Employee morale and teamwork will improve, and employees' job satisfaction will increase. Less supervision of job performance will be required, and the workforce will be more flexible and familiar with the legal requirements. Management is ultimately held responsible for drivers' job performances.[2]

Motor Fleet Training

The aforementioned benefits of training should provide enough reasons to justify training drivers. Reminding drivers how vital safety is in their profession and keeping them up to date with safety standards and regulations is one way to ensure that drivers possess the knowledge they need when operating vehicles on the highways.

Safe Attitudes

Attitudes are based on the beliefs that people possess. These beliefs are hard to change but stay with individuals for most of their lives. Through training, fleet managers are trying to alter these beliefs to reflect a safety-conscious mind-set. This is not easy, but training can foster an attitude that stresses safety in the minds of the employees.

Influencing a Driver's Attitude

There are many ways to influence a person's attitude. Continual reinforcement of safety policies and procedures through various modes of communication is vital. An organization should show the importance of safety by providing additional resources employees can use to learn and participate.

Company Bulletins

Publishing a safe driver bulletin and distributing it to all employees operating or working on motor vehicles is an excellent way to promote safety. The publication can highlight all safety awards and progress made toward various safety goals. The bulletin can also provide case studies depicting other motor vehicle accidents and questions for discussions during safety meetings.

Letters of Recognition

Sending letters of recognition to drivers who satisfy safety requirements will help influence drivers' attitudes. These letters should be personalized, and you may even want to address the driver's family. Showing employees you care enough to send letters of recognition for safety may cause employees to realize how important safety is to the company and that they should value it as well.

Posters

Posters depicting various safety slogans or messages that are placed prominently on the wall in areas where employees gather enhances the quality of your total safety focus. Posters show your employees what aspects you are most concerned with or areas that have the most problems. Posters should highlight actual company employees, if possible, for the best possible effect.

Booklets

Safety topics may be best remembered if booklets covering the main points of the topic are distributed. This will help participants retain the key points of the discussion. The booklet will also serve as resource material for any employee needing a question answered.

Safety Meetings

Safety meetings or talks are a major element in a motor fleet safety program. These meetings are designed to target one specific incident that happened or an area that needs attention. Using the 5 Ps—prepare, pinpoint, personalize, picture, and prescribe—will allow a safety manager to conduct a more effective talk. These talks also give management insight into the opinions and attitudes of the employees.

- *Prepare*: think about the subject, organize and outline your talks, and practice.
- *Pinpoint*: don't try to cover too much ground; zero in on one main idea.
- *Personalize*: establish a common ground, bring it close to home, and make it personal and meaningful to your audience.
- *Picture*: Create clear mental pictures for your listeners in order to help them see what you mean; the use of visual aids is a good way to focus on safety.
- *Prescribe*: tell participants what to do in order to help them enhance the quality of the safety meetings.[2]

According to Frank Bird, Jr., coauthor of *Practical Loss Control Leadership*,[2] effective safety meetings promote participation. Meeting leaders should encourage participation in various ways, such as asking people for their observations, opinions, or reactions; giving specific assignments; asking for volunteers; and reinforcing the positive contributions that people make during the meetings.

RECOGNITION FOR SAFE DRIVING

Every successful safety program includes activities aimed at arousing employee interest in safety. There are three main types of interest sustaining activities:

- Informative activities, which provide information as well as remind and inspire employees
- Competitive activities, which include various contests and awards based on group and individual safety performance
- Activities such as award presentation ceremonies, safety banquets, and special events in which management expresses appreciation for group and individual safety efforts

Recognition for safe driving is a great tool to show employees that managers appreciate safe driving techniques. When an organization has good drivers, the drivers need to feel appreciated for their actions not only by the company but also by the community. The following are some effective ways of relaying this message to your drivers.

Lapel Pins

Personalized lapel pins can be given to drivers as awards or as recognition of achievement. They can also be distributed easily to numerous people commemorating an achievement and are relatively inexpensive to purchase and to buy in quantity.

Monetary Awards

This is the most effective way to attract active participation in any safety recognition program. This approach creates a competitive environment in which employees can vie for cash rewards. People tend to be serious about opportunities to win money, and the competition should create a safety-conscious mind-set.

Merchandise

Giving merchandise, such as apparel, hats, bags, and golf equipment, is also an option for rewarding safe driving. This approach seems to have a positive effect on the competition among employees, and this creates an environment in which safety is emphasized.

Appreciation

Sometimes the greatest motivator is a simple pat on the back. Fleet safety directors should thank people for a job well done. A company is more successful when it employs good and productive people. A reminder that the fleet safety director appreciates driver efforts will go a long way.

ELEMENTS IN AN INCENTIVE PROGRAM

An incentive program involves the following steps:

1. *Set a standard of performance*: Tell the employees involved in the incentive program the rules of the game. Explain what poor behavior is being targeted and why it is important to limit or stop this behavior. The standard of performance established should be reachable. To win the game, the employee must be conscious of the goal every day.
2. *Show them how*: Explain to the participants the ways in which they may be able to reach the desired outcome. Show the participants the effectiveness of different ways to tackle the major anticipated problems.
3. *Keep score*: Tally scores daily and keep detailed records of points earned. Fleet safety directors are responsible for the accuracy of recording the results to ensure the rules of the game are being followed.

MEDIA TO PROMOTE SAFETY AWARDS

Fleet safety directors can use a variety of sources to promote safety awards. The media should be contacted well in advance of the awards ceremony, and a news release should be prepared before the event. The news release should explain the contents of the awards presentation and any other interesting facts and figures.

Company Publications

Publications sent or released by an organization to get the word out regarding the awards can be used to inform all members of the organization about awards ceremonies. Usually these publications are paid for by the company.

Newspapers/Television

Local newspapers and television stations are a great way to notify people of a safety awards ceremony. Editors should be notified well in advance of the activity so there is a chance for free publicity. The news release should be completed and ready to read to whoever is taking the information for the newspaper or television station.

Radio

Radio airtime is another way to promote safety awards. The fleet safety director should work with radio personnel to develop a message that tells the public what the event is and where it will be held.

To ensure a safety awards program is well covered by the media, the editor or station manager should be notified ahead of time to that person to plan ahead for coverage of the awards ceremony. Plan the award ceremonies well in advance to help prevent any miscommunications on the day of the ceremony.

ADVERTISING A SAFETY AWARDS CEREMONY

Many companies stress safety in their advertising. Remember that publicity is free but advertising must be purchased. Newspaper displays can be bought to announce awards ceremonies, but know how much space is needed (or can be afforded) and then work with it. Use the statements of various individuals in advertisements for a safety awards ceremony. Here are some examples:

- *National Safety Council*: A statement published by the National Safety Council is credible and can be used in an advertisement.[3]
- *Police department*: The authorities can emphasize the need for safety and the scope of the award.
- *Guest speaker*: An expert in a specific area unique to your organization is a good choice.
- *Insurance carriers*: Insurance companies often have individuals who specialize in motor carrier safety.

The key is to try to reach as many people as possible using the allotted budget. The budget should correspond to the key points of the event, so with careful planning and organizing it is possible to have a successful awards banquet.

STUDY QUESTIONS

1. (True or False) The costs of training drivers will be more than offset by the benefits.
2. Which is an effective way of influencing a person's attitude?
   a. Safe driver magazine
   b. Booklet
   c. Poster
   d. All of the above
3. Name the 5 Ps of an effective safety presentation.
4. (True or False) Safe attitudes are the beliefs that people possess.
5. (True or False) Monetary awards are the most effective way to get active participation in any safety recognition program.

REFERENCES

1. Della-Giustina, D.E., Profile of a good driver, *Motor Fleet Safety*, West Virginia University, Morgantown, February 20, 2002.
2. Bird, F., Jr. and Germain, G.L., *Practical Loss Control Leadership*, 4th ed., International Loss Control Institute, Loganville, Georgia, 201, 205–249, 1996.
3. Brodbeck, J.E., ed., *Motor Fleet Safety Manual*, 4th ed., National Safety Council, Itasca, IL, 176, 178, 1996.
4. Wisconsin Department of Transportation. 2010. Facilities Development Manual (FDM). Retrieved http://roadwaystandards. dot.wi.gov/standards/fdm/.

13 School Bus Safety

TRANSPORTING OUR MOST PRECIOUS CARGO

In 2008 over 24 million elementary and secondary students in the United States were bused to school daily. National figures show that approximately 68 percent of students ride a bus to school each day. Although the fatality/injury record and accident experiences for school vehicles are the lowest in the mass transportation category, the need still exists for further accident reduction.

There is no safer way to transport a child than in a school bus. Fatal crashes involving school bus occupants are extremely rare events, even though school buses serve daily in every community. Every school day, some 450,000 yellow school buses transport more than 24 million children to and from schools and school-related activities. Said another way to give perspective to the huge magnitude of pupil transportation, the equivalent populations of Florida, Massachusetts, and Oregon ride on a school bus twice every day.

In 2004–2005, the most recent year for which statistics are compiled, 55.3 percent of the over 45 million children enrolled in public K–12 schools were bused at public expense. The percentage of children bused has been declining steadily since the mid-1980s when slightly more than 60 percent were bused. At that time, the average expenditure per student transported was under $300.[1] Currently the United States spends $17.5 billion per year on school bus transportation at an average cost of $692 per child transported.[1]

Pupil transportation has become an integral part of our transportation system. The inception of Standard 17 (June 5, 1972), as one of the key areas of the National Highway Traffic Safety Administration, was designed to improve state programs for safe transportation of students.[2,3] The purpose of the standard was to reduce, to the greatest extent possible, the danger of death or injury to students while being transported to and from school. Standard 17 recommended that states expand the standard in the future to cover all youth transportation not under the jurisdiction of the Department of Transportation's Bureau of Motor Carrier Safety.[3]

According to some safety experts, the low accident fatality rate of school vehicles is not due to careful planning alone but also to the willingness of other vehicles to yield the right-of-way. Safety problems vary from school district to school district, but the safety record has been excellent during the past 20 years.

This chapter covers a number of issues relative to pupil transportation. School buses will remain a safe and efficient means of transportation with the continuing support of parents (legal guardians), school administrators, teachers, city and state officials, law enforcement agencies, and the students themselves. All of them must have a complete understanding of the school bus transportation system with a spirit of cooperation in order to conduct a successful program.

INSPECTIONS AND MAINTENANCE OF SCHOOL BUSES

If a school bus is found to be unsafe, the bus operator should discontinue its use. Such school bus vehicles, by permission of the inspecting officer, may be driven to the nearest school bus maintenance center for repair.[4] The bus may be inspected at any time by a state police officer or qualified state inspector. This is a practice carried out by all state departments of transportation. If the inspecting officer declares the vehicle unsafe for movement on the highway under its own power, the vehicle should be properly towed to the service center or other designated place for repair.

State inspections vary for all vehicles used to transport pupils to and from school or school-related events. Most states require a minimum of at least one annual inspection; however, there are states that require inspections quarterly. These inspections should be made by qualified persons employed by the state. Such inspections will not replace the daily inspections of school buses by school bus operators, the regularly scheduled preventive maintenance inspections by maintenance personnel, or the annual inspection of all motor vehicles required by the commissioner of motor vehicles in each state. These inspections should be scheduled under the supervision of the state department of motor vehicles at times and places that best protect the safety and welfare of transporting students. At the time of inspection, bus operators should present to the inspector a valid commercial driver's license (CDL), and in some states, a first-aid certificate, and a certification card issued by the state and department of education. These policies differ from state to state.

To be approved, a vehicle must meet all applicable federal and state laws, standards, and requirements. Any vehicle not meeting these requirements should be provided with an appropriate rejection sticker and may not be used for transporting students until it has been properly repaired, reinspected, and approved by the supervisor of safety.

INSPECTION OF NEW VEHICLES

In most states, it is the responsibility of the school transportation director or person designated by the county superintendent of schools to inspect each new vehicle immediately following delivery from the manufacturer. Such inspections ensure that all applicable federal and state requirements, and any specific item set forth in the purchase contract, have been met. If there are any exceptions, they must be approved by the division director of school transportation. The inspection must be completed and a valid department of motor vehicles (DMV) inspection sticker must be affixed by a qualified school bus inspector before placing the vehicle in service.[4]

DAILY AND PRETRIP INSPECTIONS

Inspections are normally performed prior to morning and afternoon trips or before any other assigned trips (such as extracurricular activities) as determined by the specific state or local county. This daily inspection is done by the school bus operator. According to the Federal Motor Carrier Safety Regulations Part 396.11, the inspection consists of examining the following:

- Service brakes
- Parking (hand brake)
- Steering mechanism
- Lighting devices and reflectors
- Tires
- Horn
- Windshield wipers
- Rearview mirrors
- Coupling devices
- Wheels and rims
- Emergency equipment

At the completion of each run, the driver should check for broken windows, ripped seats, or other damage that should be repaired before the bus is used again. Check on and under the seats for sleeping students, books, clothes, or other materials that may have been left by the students.

Housekeeping Practices and Appearance

Students should help keep the bus clean. All buses should be maintained in a safe and clean condition prior to service. It is the responsibility of the school district or contractor to provide the necessary facilities, equipment, and supplies. Cleaning should be done weekly or when necessary and should include mopping the floors, cleaning the interior upholstery, and exterior cleaning. Some school districts have their own vehicle-washing equipment.

MAINTENANCE OF SCHOOL BUSES

Many of today's buses contain sophisticated equipment such as GPS-monitoring systems, video surveillance cameras, and child-monitoring systems. These systems require a greater effort on the part of the maintenance team.

The school district or private contractor is responsible for establishing a school bus maintenance program that will ensure, insofar as possible, the safe operating condition of all vehicles used in the transportation of students to and from school and school-related events. Programs may be implemented by a school-system-operated maintenance center or under contract with a private contractor. It is to everyone's advantage that buses be designed to require minimal maintenance, including replacement and adjustment of parts and equipment.

One method for extending the life of a school bus is to replace its engine when cost effective to do so. School-system-operated maintenance centers should be staffed with mechanics and service personnel skilled in preventive maintenance and vehicle repair. The staff of school-system-operated maintenance centers should include sufficient administrative, mechanic, and service personnel to maintain and service all the vehicles in the system. Technology is being reshaped to improve safety on and around school vehicles.

The service center needs to keep the maintenance and service records of each vehicle in the fleet. A physical inventory of parts should be completed monthly. This is vital in the maintenance of vehicles because it will provide the information needed to ensure that each disabled vehicle will be back in operation as soon as possible.

RESPONSIBILITY OF PERSONNEL

Student Passengers

The school bus operator is in charge of the bus, pupils, and other passengers. School administrative and instructional personnel will cooperate with the bus operator to maintain proper discipline on the bus. The operator is to transport only those pupils enrolled in school, an employee of the school district, or a person approved by the board. Evacuation of the pupils being transported should be held at least two times a year. This instruction involves safe riding practices as well as evacuation drills. In most states, the first of these drills is to be completed by September and the second drill by early spring. Animals, weapons, and explosives are not permitted to be transported on the school bus.

Principals and teachers shall aid in instructing pupils in rules and regulations. Emphasis should also be given to pupil discipline on extracurricular trips. Passengers may not occupy any position that will interfere with the operator's vision to the front or sides, in the mirrors, or any other unsafe position.

Loading and Unloading

Loading and unloading the bus is one of the most important actions a school bus driver performs. A school bus operator must devote 100 percent of his or her attention while on the job to operating the school bus and, therefore, should not be distracted by the conduct of the passengers. Students should be instructed to behave properly when entering, leaving, and riding on the bus. Everyone has a part to play in assisting the driver to maintain order—the classroom teacher, parents, bus patrols, and last but not least, the students themselves. All school bus stops should be located safely out of the traffic stream and in areas where roads are accessible at all times. High priority should be given to the safety of pupil-passengers. The minimum sight distance should be related to the approach speed of traffic. The approach speed is the posted speed limit, advisory speed, or a value judged to most accurately represent the prevailing speed at any given location. Drivers must approach the designated stop cautiously and anticipate sudden unexpected movements at any time. The driver must always use the appropriate loading/unloading light system during the loading/unloading process in accordance with state, province, and local laws and regulations. Passengers waiting for the bus should have an area large enough to allow the group assembled to be safely away from the traffic lane.

Whenever the bus driver, parent, or school district feels the bus stop is unsafe, it may be wise to solicit cooperation from the parents, safety personnel, transportation director, bus operator, or members of the law enforcement agency to select the best available bus stop site. Reasonable efforts should be made to minimize the distance to be walked by students along narrow, heavily traveled roads without berms. If conditions develop at the bus stop site or on the bus route that compromise the safety of pupils, the bus operator should promptly bring those conditions to the attention of the transportation director. The best practical method for improvement should be implemented immediately.

EMERGENCY EVACUATION

In certain crash and emergency situations, the driver may need to evacuate the students from the bus. It is important for the driver to know when to evacuate the bus and the procedures for doing so. The bus driver is the leader and is in charge. So that students know what to expect during an evacuation, state law requires each student who is transported in a school bus to participate in emergency evacuation drills. It is recommended that the driver enlist the assistance of a student to conduct the drill. In the event of an emergency, such as an accident, fire, or the driver's inability to function, the students' lives may be endangered if they remain on the bus. In implementing an emergency school bus evacuation drill, include the following:[4]

- Policies and procedures to follow when the school bus driver is in charge.
- Procedures when the driver is unable to function.
- Instructions in the use of a cellular phone or a nearby telephone to call 911.
- Procedures in case of fire or suspected fire.
- Evacuation process: Students evacuate from either the front or rear doors depending on the situation. The escape hatches or emergency window exits might be used if exits are blocked.
- Procedures once the bus is evacuated: Students evacuated should stand at least 75 to 100 feet from the bus and off the roadway to allow emergency agencies to respond.
- Instructions covering provisions during evacuation for those students who are physically disabled. It is suggested that older, more mature, and physically stronger students be assigned the responsibility to aid in these students' safe evacuation.

It is important that a school bus emergency evacuation plan be developed, implemented, and known to all student-riders on the bus. A plan, however, is relatively ineffective unless simulated drills are conducted regularly. These same procedures should also be known to students being transported for extracurricular activities.

EMERGENCY EQUIPMENT

In case of an emergency, which may require the bus to stop in a roadway for any length of time, the operator should display warning devices such as approved bidirectional reflective triangles. The bus-flashing-hazard (four-way) lights should also be used as an additional warning to motorists. Tire chains where applicable should be carried on the bus at all times during that part of the year when snow or ice could be encountered. Bus operators must be trained in the installation and use of chains. A fire extinguisher, emergency triangles, body fluid kits, and first-aid kit should be kept in good maintenance and be easily accessible by the operators.

REPORTING ACCIDENTS

All daily, weekly, monthly, and other required reports should be complete, accurate, and promptly filed with the state department of transportation. Transportation directors should forward all reports to the state transportation director no later than 10 working days following the last day of the school month. (States have their own requirements.) An operator needs to report any road hazards to the director of transportation as soon as possible after observing or encountering them. Follow-up reports should be made to ascertain whether the hazards were corrected.

If the school bus bumps, touches, or scrapes another vehicle or object and causes damage, this is considered an accident, and it must be reported. If the school bus makes contact with a person, this is considered an accident and it must be reported. A verbal report should be made immediately, and a written report must be made within 24 hours or sooner if required by your state to the transportation director of any and all accidents in which the bus or passengers have been involved.

All major accidents involving bodily injury, fatality, extensive property damage, and or structural damage to the bus should be reported by the director of transportation to the state director of school transportation immediately, by phone, with a written report to follow within 1 week or as required by your state. All other accidents should be reported on a monthly basis to the state director of school transportation. (Most states have their own regulations.)

REGULATIONS FOR TRANSPORTED PUPILS

Outside the bus, pupils will:

- Walk on the left side of road, facing traffic.

- Be at the designated bus stop at the scheduled time for bus arrival. Parents should instruct students to wait for the bus on the proper side of the roadway, except where it would compromise their safety.
- Never stand or play on the roadway while waiting.
- Line up in an orderly fashion, safely away from the traffic lane until the bus has completely stopped.
- Board the bus in a safe, orderly manner without pushing and shoving.
- Follow the bus operator's instructions carefully by proceeding safely and alertly when getting off the bus and crossing the roadway.
- Go home promptly after alighting from the bus at the end of the school day.
- Keep the aisle clear at all times. Do not block the aisle.
- Never tamper with or block emergency exits.
- Remain silent while the bus is stopped at railroad crossings.

Once on the bus, pupils will:

- Go immediately to their assigned seats as directed by the bus operator and courteously share seats with others on the bus. (Not all school districts preassign seats.)
- Be held responsible for vandalism that occurs to the seats in which they ride. (Many school districts install cameras on their buses.)
- Change seats only with the bus operator's permission and only when the bus is not in motion.
- Get on or off the bus only when it is completely stopped.
- Cooperate with the bus operator to keep the bus clean. Eating and drinking on the bus are prohibited, except when medically necessary.
- Conduct themselves well, with quiet conversation, to enable the bus operator to give attention to safe driving.
- Avoid unnecessary conversation with the bus operator.
- Keep head and limbs inside bus windows at all times.
- Report any open exit or released latch to the bus operator immediately.
- Provide enrollment information to the bus operator as requested.
- Comply with instructions of the aide (when an aide has been assigned).

TRANSPORTATION OF DISABLED STUDENTS

Responsibility of a Public Agency

The school district or public agency will ensure that appropriate safety measures are followed in the transportation of students with disabilities. The recommended time in transit for students with disabilities should not be greater than 30 minutes beyond the time in transit of nondisabled students similarly located. Time in transit beyond this parameter should be specified in the individualized educational plan. When transportation of a student with a disability necessitates transfer while en route, appropriate supervision at the point of transfer remains the responsibility of the public agency or school district.[5]

The agency or school district will also determine the type of vehicle used to transport students with disabilities on the basis of the disabling conditions of those students. Specially adapted seats and support or protective devices will be provided for all students who require such devices to ensure their safe transportation. The public agency will terminate transportation service if parents fail to assume the responsibility of meeting the bus at the designated bus stop. Parents should be afforded due process procedures.

Responsibility of a School Bus Operator

The operator of the bus transporting students with disabilities will ensure that students aboard the bus are supervised at all times. The aide or bus operator will assist such students on and off the bus at the designated bus stop. The bus operator should verify that the protective safety devices are utilized. The bus operator and aide will receive training regarding the needs of students with disabilities. Minimal training includes the successful completion of a recognized first-aid training program.[4] For bus operators, minimal training involves the successful completion of a recognized first-aid training and cardiopulmonary resuscitation (CPR) program. This training can be conducted by the American Red Cross or the American Heart Association, or be a part of the National Safety Council's accredited programs.

Parents' Responsibilities

Parents should provide the school district or public agency with written documentation regarding any special care the student may need while on the bus. Parents are responsible for having the student at the designated bus stop at the regularly scheduled time and for providing the necessary supervision until the bus arrives. Parents are responsible for meeting the bus upon its return to the designated bus stop at the scheduled time. If a student is unable to attend school, the parents should make a reasonable and timely effort to notify the bus operator prior to the beginning of the morning bus schedule. In some states, the students are transported from their home to the school, sometimes referred as "door-to-door" pickup.[5]

EMPLOYMENT QUALIFICATIONS

No person should be employed by any board of education or private contactor unless he or she has met all contract, local, state, and federal requirements. These requirements vary from state to state. Once all clearances are in place, most states require final approval by the school district. School bus operator certification in most states is obtained by meeting the specified age, experience, license, residency, and training criteria.

States vary in the amount of previous driving experienced required to obtain a school bus license The required training and written and driving tests may be taken while the applicant is in possession of a valid commercial driver's license instruction permit. However, a valid commercial driver's license with the appropriate endorsements is required for any person who wants to operate a school bus to and from school or school-related events.

Most states require that the written test by administered by the department of motor vehicles or another state agency. The driving skills test can be administered by the state or third party examiner trained by the state. School bus certification will be granted only upon successful completion of all written and driving skills testing. Many states required recertification after a certain number of years. Any additional driving tests must conform to applicable federal regulations. Any candidate for employment as a school bus operator must receive adequate training to understand fully and carry out all the duties and responsibilities of a school bus operator.

Minimum school bus operator instruction may include and varies by state:[3]

- Classroom training
- Sufficient "behind-the-wheel" time
- First-aid training, CPR
- Special training for the transporting of children with disabilities

A record of training for each bus operator should be maintained by the state or local transportation director or school bus contractor.

EXTRACURRICULAR TRIPS

The transportation director should receive a copy of an approved schedule far enough in advance to arrange safe and adequate transportation. Schedules for approved trips should not interfere with the regular transportation schedule. Only qualified and trained bus operators should operate buses on such trips. Pupils transported by a school bus on such trips will be supervised by at least one professional employee, in addition to the school bus operator. A list of all persons on the bus should be given to the bus operator. Each additional bus should be supervised by a professional employee or person approved by the school district.

School districts providing curricular and extracurricular transportation should file, at the end of each month, a separate financial and statistical report for these trips. The reports should be filed on forms provided by the state director of school transportation. (States have their own forms for these procedures.)

STUDY QUESTIONS

1. List five pretrip inspection checks carried out by the bus operator.
2. (True or False) Students (pedestrians) should walk on the right side of the roadway with traffic.
3. (True or False) Completed maintenance and service records of all school buses in the fleet should be maintained at the service center.
4. (True or False) Administrators and teachers should aid in instructing students in rules and regulations governing students being transported on school buses.
5. (True or False) Students may transport their cats or dogs on the bus with their parents' consent.

REFERENCES

1. *Digest of Education Statistics, 2007*, U.S. Department of Education, National Center for Educations Statistics, 2008.
2. U.S. Department of Transportation, Office of Public Affairs, Washington, D.C., 2002.
3. *Federal Motor Fleet Regulations*, National Highway Traffic Safety Administration, Division of Pupil Transportation, 2000.
4. West Virginia Department of Transportation, 2001.
5. Della-Giustina, D.E., *A National Survey to Identify Current and Recommended Practices in the Transportation of Handicapped Students*, PhD diss., Michigan State University, East Lansing, 1973.

14 Shipping and Storage of Hazardous Materials

BACKGROUND

If your company transports hazardous materials, you must know the rules and regulations surrounding their shipment. The world has changed since the terrorist attacks of September 11, 2001, and the requirements for transporting hazardous materials have changed as well. There is a greater focus on hazardous materials transportation, and penalties for noncompliance can be severe. In this chapter, we examine hazardous materials transportation. It is impossible to cover everything in a single chapter, but we cover many of the highlights and provide resources for additional information.

Employees in the transportation industry are responsible for transporting just about everything. Many of those items and materials are considered hazardous materials, or *hazmat*. These items can be anything from exotic chemicals to common items such as the matches, lighter fluid, and charcoal used for barbecues.

Transportation of hazardous materials is controlled by the U.S. Department of Transportation (DOT). The DOT determines what is hazardous, in what quantities, and how it is to be packaged and handled when transported. Anything that is shipped is required to have the proper shipping papers, also called bills of lading, manifests, shipping orders, or invoices. These shipping papers must itemize the contents of the cargo since hazardous materials require special handling. When shipping the cargo of materials, all hazardous materials must be listed up front. The total quantity of hazardous material, as well as the units of measurement, must all appear on the manifest's listing.

STANDARD OF CARE

In the late 1980s a number of incidents in industry and government caused the public to have a great deal of concern. With the release of methyl isocyanate (MIC) in Bhopal, India, news of more than 2,000 deaths and 200,000–300,000 long-term and negative health issues reached people throughout the world, many of whom were concerned with "What will take place next?" A few months after Bhopal, a release of MIC also occurred in Institute, West Virginia, where MIC is manufactured. However, no fatalities resulted from this incident.

With the knowledge gained from these two incidents, the political forces across the land joined to identify the need to study and carry out research so as to formalize hazardous materials planning and response requirements. Congress and other federal agencies embarked on a number of studies designed to identify the level of preparedness that existed for chemical emergency programs at both the state and local levels. This first level of awareness covers the understanding of the use of the U.S. Department of Transportation's *Emergency Response Guide*. This guide represents the most vital piece of information available when you respond to a hazardous materials incident. The shipping papers contain information needed to identify the materials involved. All shipping papers must be kept in the cab of the motor vehicle at all times. The rules for shipping hazardous material in the United States are promulgated by the U.S. Department of Transportation (DOT) and contained in Title 49 of the Code of Federal Regulations (CFR). There are seven volumes to Title 49. Volume 2, Parts 100–185, contains the current rules and regulations concerning hazardous materials transportation.[1]

The DOT has a special group that administers the hazardous materials regulations: Research and Special Programs Administration (RSPA). RSPA develops and interprets standards, provides training assistance, monitors incidents and incident reporting, fines and penalizes offenders of the regulations, and performs inspections of carriers and shippers. In the United States, its authority reaches to all aspects of transportation, including air, ground, and water.

REGISTRATION

A shipper or carrier of hazardous materials may be required to register as a hazardous materials shipper with the DOT. A company is required to register if it offers for transportation or transport:

1. A highway route-controlled quantity of a class 7 (radioactive) material. Refer to 49 CFR 173.403 to determine if you have a highway-controlled quantity of hazardous materials.
2. More than 25 kg (55 lbs.) of a Division 1.1, 1.2, or 1.3 (explosive) material in a motor vehicle, railcar, or freight container. (See 49 CFR 173.50.)
3. More than 1 L (1.06 quarts) per package of a material extremely toxic by inhalation. (See 49 CFR 171.8 for a definition.)

4. A shipment of a quantity of hazardous material in a bulk packaging having a capacity equal to or greater than 13,248 L (3,500 gals.) for liquids or gases or more than 13.24 m^3 (468 ft^3) for solids.
5. A shipment in other than a bulk packaging of 2,268 kg (5,000 lbs.) gross weight or more than one class of hazardous materials for which placarding is required for that class.
6. A quantity of hazardous material that requires placarding. (See 49 CFR 172.500.)

Registration is done by using DOT Form F 5800.2, and a carrier must submit a complete and accurate registration statement before June 30 of each year.

Immediate Notification and Spill Reporting Requirements

All carriers are required by the DOT to submit a written report on all hazardous material incidents. The written report must be received by the DOT within 30 days from the date of the hazardous material spill. The form to be completed is DOT Form F 5800.1. Copies can be obtained from the DOT electronically @ http://hazmat.dot.gov.

The DOT requires immediate notification if the hazardous material spill results in any of the following:

- A fatality
- An injury requiring hospitalization
- Estimated carrier costs or other property damages that exceed $50,000
- An evacuation of the general public for an hour or more
- The closure of (or slowdown of traffic on) any major roadway or transportation facility for an hour or more
- The alteration of an operational flight pattern or routine of an aircraft
- Fire, breakage, spillage, or suspected contamination that involves radioactive material or etiologic agent
- A release of a marine pollutant exceeding 119 gals. or 882 lbs.
- A situation of such a nature that in the judgment of the carrier it should be reported

HAZARD CLASSES

This section gives an overview of the DOT's classification of hazardous materials. The ten classifications are as follows:

- Class 1: Explosives
- Class 2: Gases
- Class 3: Flammable Liquids
- Class 4: Solids
- Class 5: Oxidizers and Organic Peroxides
- Class 6: Toxic and Infectious Substances
- Class 7: Radioactive Materials

- Class 8: Corrosives
- Class 9: Miscellaneous
- ORM-D: Other Regulated Materials

Hazardous materials can have more than one hazard associated with one material. For example, a material such as oxygen can be both a nonflammable and an oxidizer. The following gives a brief overview of each hazard class.

Class 1: Explosives

An explosive is any substance or device that is designed to function by explosion, such as the extremely rapid release of gas and heat. Class 1 explosives are assigned a division and a compatibility group. The class, division, and compatibility group must be displayed on the hazard label. All explosives, except ammunition, must display an EX number on the package or shipping paper. The EX number is a product code that has been assigned by the associate administrator for hazardous materials. This verifies that the explosive has been appropriately classified in the correct division and compatibility group.

Definitions, classifications, and packaging requirements for explosives can be found starting at 49 CFR 173.50.[1]

Class 2: Gases

Class 2, gases, exists when a container is under a certain amount of pressure. This does not include helium balloons, carbonated pressure, or balls used for sports or tires when inflated not greater than their rated inflation pressure. A gas means a material that has a vapor pressure greater than 300 kPA (43.5 psi) at 50°C (122°F) or is completely gaseous at 20°C (68°F) at a standard pressure of 101.3 kPa (14.7 psi). Definitions for Class 2 materials can be found in 49 CFR 173.115.

Class 3: Flammable Liquids

Class 3, flammable liquids, represents the most frequent hazardous material shipped in the industry. The flash point and boiling point are the scientific criteria used in determining whether or not a liquid is a hazardous material. The flash point is the minimum temperature at which a liquid gives off vapor within a test vessel that is ignitable near the surface of the liquid.

A flammable liquid with a flash point at or above 38°C (100°F) that does not meet the definition of any other hazard class may be reclassed as a combustible liquid. This exception applies only to transportation by highway. The term "combustible liquid" is commonly used within the transportation industry and indicates that the material has been reclassed as a combustible liquid.

Flammable liquids that have been reclassed as combustible liquids are not subject to the hazardous material regulations. No hazardous materials shipping paper would need to be completed (see Appendix A for "Combustible Liquids").

Definitions for Class 3 materials can be found in 49 CFR 173.120.1

Class 4: Solids

Class 4 materials are solids and present various hazards. Class 4 materials are divided into three divisions:

- Division 4.1 Flammable Solids
- Division 4.2 Spontaneously Combustible
- Division 4.3 Dangerous When Wet

Definitions for Class 4 materials can be found starting at 49 CFR 173.124.[1]

Class 5: Oxidizers and Organic Peroxides

Class 5 materials are divided into two divisions. Division 5.1 materials are called oxidizers. An oxidizer is a material that may, generally by yielding oxygen, cause or enhance the combustion of other materials.

Division 5.2 materials are called organic peroxides. Organic peroxide means any organic compound containing oxygen (O) in the bivalent -O-O- structure and that may be considered a derivative of hydrogen peroxide, where one or more of the hydrogen atoms have been replaced by organic radicals or other conditions apply.

Definitions for Class 5.1 and 5.2 materials can be found in 49 CFR 173.127, 173.128, and 173.129.[1]

Class 6: Toxic and Infectious Substances

Class 6 materials are divided into two divisions: Division 6.1 (toxic or poisonous) and Division 6.2 (infectious substances). A toxic material, other than a gas, is a material that is known to be so toxic to humans that it presents a hazard to health during transportation. Material toxicity may fall into three categories: oral, dermal, and inhalational. Division 6.2 are materials known to contain, or are suspected of being, a pathogen.

Class 7: Radioactive Materials

Radioactive material is any material having a specific activity greater than 70 Bq per g (0.002 microcurie per g). Refer to 49 CFR, Part 173, Subpart I, for further information regarding Class 7 materials.[1]

Radioactive materials are divided into three categories: White I, Yellow II, and Yellow III. The requirements for radioactive materials can be found starting with 49 CFR 173.401.[1]

Class 8: Corrosives

A corrosive material is a liquid or solid that causes full-thickness destruction of human skin at the site of contact within a specified period of time. Refer to 49 CFR 173.136 and 173.137 for further information.[1]

Class 9: Miscellaneous

These materials do not meet the definition of any of the other hazardous classes. Miscellaneous hazardous material (Class 9) are materials that present a hazard during transportation but that do not meet the definition of any other hazard class. This class includes:[2]

- Any material that has an anesthetic, noxious, or other similar property that could cause extreme annoyance or discomfort to the operator of a vehicle
- Any material that meets the definition for an elevated temperature material, a hazardous substance, a hazardous waste, or a marine pollutant

The terms "hazardous substance" and "hazardous waste" are used to identify those materials that are also regulated by the Environmental Protection Agency (EPA).

A hazardous material that is also a hazardous substance is identified by the letters RQ (Reportable Quantity). These materials are considered hazardous material during transportation only because they are regulated by the EPA due to the harm they could cause to the environment if released.

ORM-D: Other Regulated Materials

An ORM-D material (although regulated as a hazardous material) presents a limited hazard during transportation due to its form, design, quantity, or packaging. The regulations concerning the transportation of ORM-D materials via highway (ground) versus aircraft (air) are very different. Transportation of ORM-D materials by highway does not require shipping papers. The majority of ORM-D materials are consumer commodity products but can include ammunition (cartridges, small arms, and cartridge power device). It is a shipper's responsibility to classify its product as an ORM-D material. Please refer to the ORM-D classification section for further details.

HAZMAT EMPLOYEES AND EMPLOYERS

All employees involved in the hazardous materials (in this section referred to as hazmat) transportation process must be trained. It is the employer's responsibility to make sure all employees are trained. The DOT outlines the following definitions for a hazmat employee and employer:

A hazmat employee is a person who is employed by a hazmat employer and directly affects hazmat transportation safety. This includes an owner/operator of a motor vehicle that transports hazmat. It also includes a person (even a self-employed person) who:

- Loads, unloads, or handles hazmat
- Tests, reconditions, repairs, modifies, marks, or otherwise represents packaging as qualified for use in the transportation of hazmat
- Prepares hazmat for transportation

- Is responsible for safety of transporting hazmat
- Operates a vehicle used to transport hazmat

A hazmat employer is a person or company that uses one or more of its employees in connection with:

- Transporting hazmat in commerce
- Causing hazmat to be transported or shipped in commerce
- Representing, marking, certifying, selling, offering, reconditioning, testing, repairing, or modifying packaging as qualified for use in the transportation of hazmat

A hazmat employer also includes any department, agency, or instrumentality of the United States, a state, a political subdivision of a state, or a Native American tribe engaged in offering or transporting hazmat in commerce.

HAZMAT TRAINING

Hazmat employee training should cover the methods and observations that may be used to detect the presence or release of a hazardous chemical in the work area. These methods include monitoring conducted by the employer, continuous monitoring systems, visual appearance, or odor of hazardous chemicals being released, as well as the physical and health hazards of any chemicals in the work area.

The training should also address the measures employees can take to protect themselves from these hazards, including specific procedures the employer has implemented to protect employees from exposure to hazardous chemicals, such as appropriate work practices, emergency procedures, and personal protective equipment (PPE) to be used.

The employee should also know the details of the hazard communication program developed by the employer, including an explanation of the labeling system and the material safety data sheet, and where employees can obtain and use the appropriate information. Initial training must be conducted for all employees. New employees must be trained prior to their initial assignments. The training of a hazmat employee must include:

- *General awareness and familiarization training*: Each employee should be provided with training in respect to the hazmat requirements. This training must enable the employee to recognize and identify hazardous materials.
- *Function-specific training*: Each employee must be provided training applicable to the functions the employee performs.
- *Safety training*: Each employee must receive safety training that includes emergency response information, measures to protect the employee from the hazards associated with the material, and methods and procedures for avoiding accidents such as safe handling of packages.

A new hazmat employee (or an employee who changes job functions) may perform those functions prior to the completion of the training provided the employee performs those functions under the direct supervision of a properly trained and knowledgeable hazmat employee. In addition, the training must be completed within 90 days of employment or a change in job function.

A hazmat employee should receive the training at least once every 3 years.

Employees must receive additional training whenever:

- New hazardous substances are introduced into the workplace.
- Exposures to hazardous chemical change.
- Employees are subject to increased exposure due to changes in work practices, processes, or equipment.
- Additional information about the hazardous substance in the workplace becomes available.

SHIPPING PAPERS

A shipping paper is required by the DOT when a company is transporting a hazardous material. The shipping paper is used to describe the material as well as provide appropriate emergency response information in the event of an incident to those individuals involved in handling and transporting the package. The shipping paper must be accurate and complete.

The DOT requires that the shipping paper include the basic description in a specified sequence. The new specific sequence will be mandated in 2013. The DOT basic description consists of the following:

- The DOT proper shipping name
- Hazard class or division number
- Identification number
- Packing group
- Weight

The basic description cannot include unnecessary information, such as product codes or inaccurate and incomplete information (See Appendix B "Describing Hazard Materials.")

EMERGENCY RESPONSE

The *Emergency Response Guidebook* (*ERG 2000*) was developed jointly by the DOT, Transport Canada, and the Secretariat of Communications and Transportation (SCT) of Mexico for use by firefighters, police, and other emergency services personnel who may be the first to arrive at the scene of a transportation incident involving a hazardous materia1.[2] It is primarily a guide to aid first responders in (1) quickly identifying the specific or generic classification of the material(s) involved in the incident, and (2) protecting themselves and the general public during this initial response phase of the incident. The DOT will allow exceptions for certain materials, such as those shipped under the limited quantity exceptions. The *ERG* is updated every 3 to 4 years to accommodate new products and technology. The next version is scheduled for 2012.

CONTACTING FIRST RESPONDERS

Emergency contact telephone number must be provided. The number can be toll-free or a local number. Local numbers must include the area code in which they are located. The number must be available 24 hours and be capable of providing emergency response information at any time during transportation. The emergency contact number is used for emergency response information. It is used when there is an incident involving the hazardous material. Please refer to the *ERG 2000* for further details.

PLACARDING

When shipping hazardous materials, the vehicle may be required to be placarded. Placards must be placed on each end and side of a transport vehicle, railcar, bulk packaging, and so forth. Certain materials require that regardless of quantity shipped the vehicle must be placarded. Other materials require that if there are over 1,001 lbs., then the vehicle must be placarded. This is determined by using the two placarding tables in 49 CFR 172.504(e).[3]

SECURITY

In the wake of the terrorist attacks of September 11, 2001, and subsequent threats related to biological and other hazardous materials, the DOT has mandated that all carriers and shippers implement a security awareness program. A security plan is also required in certain instances (see 49 CFR 172.800). The security plan must address the following:[1]

- Personnel security
- Unauthorized access
- En route security

HAZARD COMMUNICATION

The purpose of a hazard communication program is to comply with the Occupational Safety and Health Administration (OSHA) Hazard Communication Standard 29 CFR 1910.1200 and to warn employees about hazardous chemicals and substances in the workplace.[3]

A written program is required to achieve compliance with the requirements of the standard. The specific methods described in this written program are for illustrative purposes, and other effective methods may be substituted to satisfy local needs or practices.

Hazmat employers should provide employees with information and training on hazardous chemicals in their work area at the time of their initial assignments. All employees also need to receive additional training when a new hazard is introduced into the work area.[2]

The employees should be informed of any operations in their work areas where hazardous chemicals are present, and they should know the location and availability of the written hazard communication program. This includes the list(s) of hazardous chemicals and material safety data sheets required by this section.

CHEMICAL INVENTORY

A hazmat business should determine what hazardous materials it uses and maintain a current list of the kinds and amounts.[4] The location of these materials should also be noted to aid in an emergency situation. Material safety data sheets (MSDSs) should be obtained and kept on hand for reference as well as for employee training.

All hazardous substances must be stored and labeled properly. All containers, no matter how minute the quantity, should be compatible with the material they contain and properly labeled to prevent accidental misuse. The National Fire Protection Agency (NFPA) Codes and 29 CFR 1910 dictate the proper storage, grounding, and dispensing of hazardous materials. Dangerous chemical reactions can result if certain substances are mixed together.[3]

Other chemicals, such as paints, thinners, solvents, and fuels, have specific storage requirements that should be carefully followed. Temperature, ventilation, vibration, and close proximity to other substances can have adverse effects on certain hazardous materials.

Transferring of chemicals should be carefully done according to manufacturers' procedures, with containers grounded to prevent fire or explosion from static electric charges. The NFPA has a label known as the "704 diamond" that informs users at a glance of the properties of a substance.

PERSONAL PROTECTIVE EQUIPMENT

All necessary site-specific personal protective equipment (PPE) mandated for use within a hazmat company must be used to prevent hazardous materials from harming personnel, property, and the environment. Records of issuance must be maintained, and the company should train its employees in the use and maintenance of the PPE.

A program to train new employees and periodically update training of current employees should be in place. Any medical screening and records, such as a pulmonary function test for respirator wearers, should be maintained, and deficiencies must be noted for correction.

The manager should make sure that each employee is using his or her PPE properly and is not disabling or bypassing engineered safeguards. Any outside contractors who handle hazardous materials should be required to submit written safety procedures and periodically monitored to ensure compliance.

SPILL PLAN

Any spillage of hazardous materials should be cleaned up immediately. Any damage to containers of hazardous substances should be reported, and the containers must be repaired, replaced, or otherwise contained immediately.

A viable spill plan to protect hazmat personnel, property, and the environment should be developed. Equipment for use in spill incidents should be kept on hand and properly maintained. Employees should be trained to use this equipment, and trained employees should be present during each shift.[4]

Periodic spill containment drills can be helpful in quickly correcting a potentially dangerous and costly situation. Coordination and local emergency service notification procedures should be instituted and maintained. The procedures should also include contacting the state and federal environmental protection officials. Local, state, and federal officials may want to know the type, amount, and location of any hazardous substance that your organization stores on site in order to be more effective in their response should they ever be called upon.

STUDY QUESTIONS

1. What is the group within the DOT that administers the hazardous material regulations?
2. Name the requirements for hazmat registration with the DOT.
3. What form is used to submit a hazardous materials incident report? What is the time limit by which it must be submitted?
4. Under what circumstances is immediate notification of the DOT required when a hazmat incident occurs?
5. Name the ten hazard classes of hazardous materials.
6. What are the three types of training required for all hazardous materials employees?
7. What are the requirements for initial hazmat training and recurrent training?
8. What is the basic description required on hazardous materials shipping papers?
9. Name the book that contains emergency response information for use by first responders to hazardous material incidents.
10. What is the purpose of the hazard communication standard?

REFERENCES

1. Battelle Memorial Institute, *Comparative Risks of Hazardous Materials and Non-Hazardous Materials Truck Shipment Accident/Incidents*, (retrieved April 12, 2008).
2. Bureau of Transportation Statistics, National transportation atlas-database/2007/
3. Cova, T.J. and Conger, S., Transportation hazards, in M. Kutz, ed., *Handbook of Transportation Engineering*, McGraw Hill, New York, 2–3, 2004.
4. Federal Highway Administration, *Freight Analysis Framework for Tennessee*, 2008, www.ops..fhwa.dot.gov/freight- analysis/faf/state-info/faf2/tn.htm (retrieved April 30, 2008).

5. Phelps, R., *All Hazards vs Homeland Security Planning*, 2005, http;//www.ems-solutionsinc.com/pdfs/MeettheExperts.pdf (retrieved October 30, 2008).
6. Tennessee Department of Safety, Commercial Vehicle Enforcement, *Size and Weight Limits for Trucks*, 2008, http://www.state.tn.us/eed/pdf/highway (retrieved April 11, 2008).
7. Code of Federal Regulations, Title 49, Parts 40, 100–185, 325, 355–399, U.S. Department of Labor.
8. *Emergency Response Guidebook*, U.S. Department of Transportation, Research and Special Programs Administration, Office of Hazardous Materials Training (DHM-50), Washington, D.C., 2000.
9. Code of Federal Regulations, Title 29, Part 1910, U.S. Department of Labor.
10. Della-Giustina, D.E., *Fire Safety Management Handbook*, 2nd ed., American Society of Safety Engineers, Des Plaines, Illinois, 2003.

15 Security in the Transportation Industry (Transit and Motor Carrier)

FEDERAL TRANSIT SYSTEMS

In the wake of the terrorist attacks of September 11, 2001, the U.S. Department of Transportation (DOT), through many of its agencies, launched an aggressive program to enhance the security of the nation's public transportation programs. This chapter explores the Motor Carrier and Federal Transit Administration (FTA) systems. A total of thirteen agencies are currently in operation. Two former DOT agencies were transferred over to the new Department of Homeland Security (DHS). President George W. Bush signed legislation creating the new department, whose goal is to address terrorist threats against Americans at home. Also, the U.S. Coast Guard and the Transportation Security Administration moved from the DOT to the DHS on March 1, 2003.

Transit systems are inherently open environments and are designed to move people through an urban area with easy access for passengers. In the past year, many gains in protection and preparedness have required different alternatives as well as management interagency coordination and financial investment.

PLANNING

Since the early 1990s, the nation's 100 largest rail and bus companies when combined moved approximately 85 percent of all passengers who utilize public transportation. Defending against terrorism is not new to this discipline. An effective response to an act of terrorism requires instantaneous and sound decision making in a volatile, high-pressure environment. Today, some of the largest systems have emergency response plans but need to reexamine their plans to meet potential threats. The plans can serve as blueprints for action in case of an attack. Key steps to take include notifying authorities of the incident, evacuating passengers, protecting personnel and equipment, activating a unified command and communications system among transit police and emergency fire and medical units, and restoring the system to normal operation.

As a result of the tragic events of September 11, safety personnel have emphasized how important it is to implement regular emergency response drills. Every transit agency should conduct regular emergency drills, and not just fire drills, to keep skills sharp, update response plans, and build personal relationships with counterparts in the police, fire, and emergency medical response organizations.

It is important that all members of the transit workforce understand security issues and be prepared to respond should an emergency occur. Participants will gain a better understanding of the roles played by the various agencies and begin the process of developing plans and relationships necessary to respond effectively in an emergency situation. Since the 9/11 attacks in New York, Washington, D.C., and Pennsylvania, the FTA has supported U.S. industries' programs for security and preparedness through training, research, and various regulations and has worked closely with the public transportation industry. In response to new threats of terrorist activity, FTA has implemented a national transit security program that includes the following elements:

- *Security assessments*: FTA has deployed teams composed of experts in anti- and counterterrorism, transit operations, and security and emergency planning to assess security at thirty-six transit agencies. FTA plans to extend the programs to additional agencies after the first thirty-six assessments are complete.
- *Emergency response planning*: A plan is in place to provide technical assistance to the sixty largest transit agencies for creation of security and emergency plans, together with emergency response drills.
- *Emergency response drills*: Grants have been offered for organizing and conducting emergency preparedness drills. Approximately $3.4 million has been awarded to more than eighty transit agencies through these grants.
- *Security training*: "Connecting community" forums have been designed to bring together small- and medium-sized transit agency personnel and local public safety agencies, including law enforcement, fire, and emergency medical services, and specialized units for hazardous materials and explosives disposal. These forums will provide participants with many opportunities.
- *Research and development*: FTA has increased the funding of its safety- and security-related technology research to enhance a systems approach to the problem of release detection and management, including modeling and simulation, to identify the most effective application of sensor-based technology, communication systems, training tools, and exercise and response planning.

Today there are over 7,500 local public transportation systems in the United States. One primary concern of the federal government is to encourage security and preparedness programs that will protect all of them. This might seem like an impossible task when we compare the various systems across the country. Each system has a unique method of carrying out security measures that best suits its needs. Daily in the United States, publicly funded transportation systems provide about 32 million passenger trips. They provide transportation to students, commuters, the aged, tourists, disabled individuals, and others who rely on trains, buses, vans, ferryboats, and other accessible vehicles and facilities to reach their final destination.

The public transportation industry has to cope with many threats, all of which have the potential for disrupting local communities, causing casualties, and damaging and destroying public and private property. At the national level, the industry will most likely be affected by several major disasters each year, such as earthquakes, floods, tornadoes, hazardous materials spills, major accidents, fires, violent crime, and potential terrorist acts.

Implications for Terrorism

Threat assessments issued by the FBI have consistently placed public transportation at the top of the critical infrastructure protection agenda, along with airports, nuclear power plants, dams, and major utility exchanges on the national power grid.

Terrorist acts are not accidents or disasters, but they are intentional actions designed to inflict casualties. In order for transportation systems to address these threats, they must be able to (a) recognize and prevent potential incidents and emergencies, and (b) enhance response capabilities.

Types of Training

The types of training programs that security management could provide are important and meaningful to the enterprise. As professionals, we know that inadequate training is the primary contributor to poor job performance. The training function means different things to different people and is often widely misunderstood. Under the management plan, there are elements that management wants their new employees to know when discussed during the initial training program: what to do, why the task is covered, and how they want it done. The two basic strategies to training include formal classroom training and on-the-job training. In implementing this process, the selection and the experience of the trainer is indeed a key point in security training. When organizing the program, a discussion of the complexity of any security position during the classroom phase of instruction should be included. During this particular phase, lectures and leaders in the field along with training films are appropriate methods to use in the classroom. After the on-the-job training aspect of the curriculum, it is necessary for documentation of such training to be placed in the driver's personnel training file. This file is subject to inspection in the event of a legal problem, especially when a lawsuit claims the security individual was not properly trained on his or her responsibilities on the assignment.

Training and Exercising

When evaluating the transportation system's needs for training and simulated drills, key elements to remember are:

- All training should reflect security and emergency plans and procedures.
- Simulated drills are conducted to assess the quality of both training and plans. These will provide valuable feedback in the planning process. The reason companies need training for security and emergency preparedness is to provide transportation system personnel with the specific knowledge needed to perform the critical functions required in system plans and procedures. Training in this regard may be highly technical, geared specifically to the responsibilities of an individual employee to support the system during an emergency. Throughout the country, transportation systems of all sizes provide a wide range of training for their employees, contractors, and local public safety agencies.
- Emergency preparedness is a continuous process consisting of three integral functions: program planning, employee training, and conducting simulated drills. Each function is dependent on the other two and should not be viewed in isolation.

MOTOR CARRIER SYSTEMS

Driver Safety

Motor carrier operators often work alone, at night, and in surroundings they are not familiar with, and they are responsible for valuable cargoes. Recent events have focused a great deal of attention on the vulnerability of the nation's infrastructure to major events, including terrorism. Emergency management in public transportation is constantly evolving, incorporating lessons learned from major events as well as facing new threats. Public transportation systems at the forefront of security and emergency management offer the vision needed to guide industry efforts for enhanced capabilities.

Once the driver is in transit, he or she should make sure all tractor and trailer doors are locked and the windows are rolled up. When stopped in traffic, the driver must leave enough space in front of the vehicle in order to pull away if he or she encounters trouble. When driving on expressway ramps and in urban areas, drivers are in particular danger and must be alert.

Scheduled Truck Stops

The trucking industry needs to address the issue of overturned tractor-trailers and other highway incidents. Motor carrier and transit operators need to spend a great deal of time on highway safety, focusing on the driver and the general public.

Instruct drivers to follow these safety steps:

- In planning your regular routes, vary your routine so hijackers cannot predict where you will stop. Schedule your stops in advance and avoid unplanned stops.
- Lock all doors when you park, and leave your vehicle secured by placing all valuables out of sight of strangers and onlookers.
- Avoid dark and empty parking lots. Park in a well-lit lot when possible.
- Develop a personal security plan. Follow this plan at all times.
- When reentering your vehicle, check to make sure no strangers are inside. Once inside, lock your doors.
- Standard procedure for transit operations based on the terminals where passengers are waiting to go aboard should be followed. Usually, transit operators have a list of all passengers in advance.
- Report any unusual conditions to your dispatcher.
- Because thieves may be monitoring your radio conversations, never discuss anything about the cargo you are transporting.

CHESHIRE PETROLEUM & GAS CORPORATION (MODEL PLAN)

The model motor fleet safety plan presented in this chapter is based on an estimated number of businesses and residents that are located within four states: West Virginia, Pennsylvania, Ohio, and Maryland. The total coverage is in excess of an 800-mile radius, supporting the fact that a large motor fleet department is necessary to cover such a large area. There are approximately 500 trucks to carry out the tasks of meter reading, maintenance, and delivery of their services. The primary mission of Cheshire Petroleum & Gas (a fictitious company) is to develop and implement the policies, procedures, and plans so that a motor carrier safety program under the direction of a competent fleet safety director will provide the best program. To be effective, the program should cover a variety of tasks, including an adequate training program, procedures for loading and unloading freight, highway safety procedures (focusing extensively on the driver and the general public), and proper investigation techniques. With upper management directing its attention to these areas, the operation will be much safer and more productive.

MOTOR CARRIER POLICY

The motor carrier policy for Cheshire Petroleum & Gas includes the following:

- Date that the policy was adopted.
- Objectives of the policy depicting vehicles purchased to provide the best possible support to the corporation's operations, environmental impact, and other governmental objectives and guidelines.

- A policy statement to show that the motor vehicles are used only when required to conduct company business.
- Fleet vehicles should be managed in accordance with the lifecycle approach to material management, including the principles of economy and minimizing any negative environmental impact.

The safety director has many responsibilities to cope with and is required to spend much time and resources when a potential conflict arises between DOT regulations and safety and health standards enforced by the Occupational Safety & Health Administration (OSHA). Other duties that are required include investigations; planning, organizing, and selling the program to the entire enterprise; and a follow-up reporting system.

CRASH INVESTIGATION PROGRAM

INVESTIGATIONS

Investigations must be conducted for all incidents and should include determining where the corporation should concentrate its safety and health efforts. The fleet safety director should examine past data and records, including police and insurance reports. Upon analyzing the data, the director or investigative officer should be able to identify the types of high-frequency incidents that can be addressed during safety training. In addition, the director or investigator should check work procedures, job responsibilities, and the various OSHA forms for information.

SUPERVISOR'S PROTOCOL WHEN INVESTIGATING ACCIDENTS

There are protocols for the supervisor to adhere to when conducting a motor vehicle investigation. This is a policy of Cheshire Petroleum & Gas and has precedence over other ongoing activities. All investigations are completed by a trained, experienced individual utilizing standard investigation protocols. Furthermore, all crashes of fleet vehicles require that an on-site investigation be conducted and that the vehicle involved be removed from service until preventive maintenance has been performed to ensure that mechanical damage has not occurred (which could cause another crash).

OBJECTIVES OF CRASH INVESTIGATIONS

The goals of crash investigations are:

- Secure the accident scene.
- Preserve vital information.
- Provide organizational support for injured employee(s).
- Gather useful preventive accident information for analysis.

Postaccident Interviews

The role of the crash investigator is of primary importance for Cheshire Petroleum & Gas employees and for the company itself, because the investigator should be the first representative of the organization on site. He or she should meet with law enforcement, emergency medical services, media, injured employees, and any other injured parties. The investigator must be constantly vigilant concerning the ways to control loss for the organization and shield against potential liability. The first and foremost duty is to search for all facts and be able to reconstruct the crash completely. This type of data is critical.

The crash investigator will use a prepared kit that includes the following items:

- Accident investigation reports
- Camera
- Measuring tape or other suitable measuring device and chalk
- Flashlight
- Clipboard and pen
- Cellular telephone

Procedures

The motor fleet carrier should follow these procedures:

1. When management is notified of a motor vehicle accident, the employee taking the call should contact the highest ranking supervisor to initiate the investigation.
2. The investigator should obtain the investigation kit and go to the scene.
3. After gaining an initial overview of the accident, the investigator should begin to photograph the scene from several vantage points.
4. The investigator should contact the local law enforcement agency.
5. The investigator should meet with any other involved parties and emergency medical services to offer assistance.
6. The investigator should interview any witnesses to the crash. The interviews should be made at or near the accident scene. The investigator should obtain witness contact information.
7. The investigator should accompany the employee to the medical facility, if necessary.
8. The investigator should complete all the necessary reports. The driver will be placed in nondriving status and have driving privileges revoked until the crash investigation findings are complete.

Postaccident Interview with Driver

The investigator will conduct a postaccident interview with the driver, primarily to resolve any discrepancies existing in the accident report. The driver will provide any needed explanations regarding the crash.

Telephone Checklist for Motor Vehicle Accident

When investigators call their companies' home offices to report information, they must be prepared to answer the following questions:

1. What is your name?
2. Where did the accident occur?
3. Have the police been contacted?
4. Has anyone been injured? If so, who?
5. Are hazardous materials involved?
6. What happened? (Brief summary, including names of witnesses)

The person taking the information should record the time the call was received and the date, and then sign his or her name at the bottom of the checklist.

PROGRAM DEVELOPMENT

Planning

Planning consists of forming a plan of action. This task may require Cheshire Petroleum & Gas to change certain procedures or even to provide additional training. It is important to realize that plans must be flexible and to believe in the phrase "continual improvement."

A responsibility of the fleet safety director of Cheshire Petroleum & Gas is to promote the entire safety program throughout the entire organization. The only way in which the program will be a success is if the director can integrate the system throughout every level of the company. The director may consult the company's insurance carrier to provide additional information and to help with developing a cost-benefit analysis. This information can be used to obtain upper management's support on future safety expenditures. The safety director should be able to present all of the information in a factual and businesslike manner.

Training

Training is a crucial part of the motor fleet safety program. The fleet safety director should oversee development and implementation of the job training, job specifications, and work procedures. This ensures that all legal and mandated regulations are covered. The entire program must be in place and outlined, and training procedures must be clear and concise.

Follow-Up Procedures

The follow-up programs and procedures should consist of five different aspects:

- Monthly report of absenteeism due to accidents
- Regularly scheduled inspections, conducted by a committee of Cheshire Petroleum & Gas personnel
- Return of incomplete accident reports to obtain complete information
- The recording of findings in the driver's file when remedial action is indicated
- Continual analysis of accident and injury records

By following the above procedures, the fleet safety director will help Cheshire Petroleum & Gas achieve better safety programs in the future.

Reporting

Reporting safety results is also a duty of the motor fleet safety director. This occurs at a regular time, whether it be every month or every other month. At this time, the director will report all achievements as well as problems to both management and employees. The safety director must be fair and accurate. All reports should also emphasize any problem as another learning experience.

SELECTING, TRAINING, AND SUPERVISING PERSONNEL

Driver selection is a critical component of the motor fleet safety program. The company wants to make sure that it obtains qualified and safety-conscious drivers. A good driver must be able to avoid accidents, follow traffic regulations, perform pre- and posttrip inspections, avoid abrupt starts and stops, avoid schedule delays, avoid irritating the public, perform the nondriving parts of the job, get along with others, and adapt to meet existing conditions.

Previous experience as either a professional driver or a private motorist is the best indicator of the candidate's ability to drive safely. Therefore, the human resources manager should consult with an applicant's previous employers and state license records to see the applicant's driving record. The driver must have a current state driver's license.

The degree of an applicant's driving skills can be determined by placing the applicant behind the wheel of a vehicle and observing his or her driving ability over a prescribed course. The course can be of two types. One is called a driving range test; this determines the driver's awareness of the physical dimensions and limitations of the vehicle. The second type consists of taking the driver through a traffic situation where he or she would be exposed to certain driving conditions.

All drivers will receive training from their initial hiring date to the time they leave. Training is important for reducing accidents, reducing maintenance costs, reducing absenteeism and labor turnover, lessening the burden of the supervisor, and improving public relations. All drivers will receive training through videos and lectures, as well as behind-the-wheel training. The training will cover safety procedures, how to conduct inspections, how to respond during an accident, and how to maintain records, among other topics.

DRUG AND ALCOHOL TESTING PROGRAM

Alcohol Testing

A program for alcohol testing should focus on these factors:

- The testing should be conducted mainly for employees in safety-sensitive positions (drivers, maintenance employees, etc.).
- All drivers and prospective drivers must have a commercial driver's license (CDL) before they can be tested, in accordance with the Federal Highway Administration.
- Drivers will not take part in any safety-sensitive tasks under the following circumstances:
    - While having a breath alcohol level of 0.02 percent and higher
    - While using alcohol
    - Within 4 hours of using alcohol
- Drivers may not refuse to submit to an alcohol test.
- Drivers are not permitted to consume alcohol within 8 hours after an accident or until tested.
- Drug and alcohol tests are required for all of the following situations:
    - *Preemployment*: Conducted before performing safety-sensitive functions for the first time.
    - *Postaccident*: Conducted after accidents on drivers whose performance may have contributed to the accident.
    - *Reasonable suspicion*: Conducted when the fleet safety director observes behavior that resembles alcohol misuse.
    - *Random*: Conducted on a random and unannounced basis. This test can occur before, during, and after performance of safety-sensitive tasks.
    - *Return to duty and follow-up*: Conducted on a driver who previously violated prohibited alcohol conduct standards. The procedure for this test will include six tests in a 12-month period after the driver has returned to work.
- The alcohol tests that will be conducted by the company will utilize evidential breath testing (EBT) devices that are approved for use by the National Highway Traffic Safety Administration.
- According to company policy, if an EBT test shows a breath alcohol level above 0.02 percent, then a confirmation test must be conducted on an EBT a second time. The second test prints out the results, date, time, the sequential test number, and the name and serial number of the device, and determines what actions must be taken, according to the results of the second test.
- According to company policy, a breath alcohol level above 0.02 percent requires the driver to be prohibited from driving any company trucks for a period of 24 hours. In addition, if a driver's appearance and behavior

suggest possible alcohol misuse but the EBT test cannot be conducted, it is the responsibility of the fleet safety manager to ban the driver from driving for 24 hours.
- If the fleet safety director decides to allow the driver to return to work, the fleet safety director must ensure that the driver:
  - Has been evaluated by a substance abuse professional
  - Has complied with any recommended treatment put forth by the substance abuse professional
  - Has taken a return-to-work EBT test showing a breath alcohol level below 0.02 percent
  - Understands that he or she is subject to unannounced follow-up tests
- Violations of the company-based testing rules will not affect the driver's CDL record.
- It is the fleet safety director's obligation to provide drivers with information about alcohol misuse, the company's policies, the testing requirements, and how and where drivers can get help for alcohol abuse.
- Fleet safety directors should be trained to recognize alcohol abuse; training is a key element.

Drug Testing

Drug testing is required for the same events as alcohol testing; however, a split test is required. A split test is a test in which the sample, provided by the driver in question, is divided into two equal samples:

- The primary sample will be tested by a lab that has been previously approved by the Department of Health & Human Services (DHHS).
- If the analysis of the primary sample confirms the presence of illegal, controlled substances, the driver has 72 hours to request that the split sample be sent to another previously approved DHHS lab.

These samples are analyzed only for the identification of the following substances:

- Marijuana
- Cocaine
- Amphetamines
- Opiates (including heroin)
- Phencyclidine (PCP)

A confirmation test will be conducted by an approved lab using gas chromatography/mass spectrometry (GC/MS) if the initial screening test is positive.

The positive test will then be reviewed by a National Highway Traffic Safety Administration (NHTSA) medical review officer, who will then contact the driver in question to determine if there is a legitimate medical purpose that explains why the employee is taking the drug. If there is no legitimate purpose, then the medical review officer will contact the fleet safety director for further action.

A positive test result will carry the same effect on the driver's work status as an alcohol violation.

HEALTH AND FITNESS QUALIFICATIONS FOR DRIVERS

The Cheshire Petroleum & Gas Company requires the following physical qualifications for drivers based on NHTSA standards:

- A person should not be a company commercial truck driver unless he or she is physically qualified to do so or has the original or photographic copy of a medical examiner's certificate that states he or she is physically qualified to do so.
- A person is considered physically qualified to be a company commercial truck driver if that person:
  - Has no loss of a foot, a leg, a hand, or an arm
  - Has no impairment of a hand or finger that interferes with wheel grasping and operation, or impairment of an arm or leg that interferes with the person's ability to perform normal tasks associated with operating the truck, or any other significant limb defect or limitation that interferes with the person's ability to perform normal tasks of truck operation
  - Has no established medical history or clinical diagnosis of diabetes mellitus, which requires insulin for control
  - Has no current clinical diagnosis of myocardial infarction, angina pectoris, coronary insufficiency, or any other cardiovascular disease accompanied by congestive cardiac failure
  - Has no established medical history or clinical diagnosis of a respiratory dysfunction that would interfere with the driver's ability to operate safely
  - Has no clinical diagnosis of high blood pressure
  - Has no established medical history or clinical diagnosis of arthritic, orthopedic, muscular, neuromuscular, or vascular disease
  - Has no established medical history or clinical diagnosis of epilepsy
  - Has no mental, nervous, or functional disorder or psychiatric disorder
  - Has at least 20/40 vision
  - First perceives a whispered voice no less than five feet without the use of a hearing aid
  - Does not use Schedule I drugs such as:
    - Amphetamines
    - Narcotics
    - Any habit-forming drugs, unless prescribed by a medical practitioner
- Has no clinical diagnosis of alcoholism

Drivers Should Be Tested for Compliance

Driver supervision will be carried out by the fleet safety director. These tasks include:

- Personally observing the driver's performance

- Checking driver trainer reports, road patrol reports, arrest records, and comments from other drivers and employees
- Reviewing complaints from other drivers or pedestrians as a possible barometer of driving performance
- Being alert to personality and performance changes before they reach the incident stage by making frequent personal contacts with the drivers

The fleet safety director should watch for the following typical symptoms of accidents in the making: errors in performance of work, changes in everyday behavior and manners, changes in simple habits or routines, and near accidents. Picking up on these warning signs will enable the supervisor to make changes before an accident occurs.

PRETRIP/POSTTRIP INSPECTIONS

The maintenance and safety departments must work together to devise an inspection plan. The driver needs to be aware that he or she must conduct these inspections and know how to conduct them and what to look for, as well as keep an accurate record of his or her findings and a daily statement. A mechanic can use the driver's records to find out what is wrong with the vehicle, and the driver can check to see what has been done to the vehicle.

Pretrip Inspection

The driver must be sure that the vehicle is safe for travel. Fifteen minutes should provide ample time to check the DOT-required components likely to cause safety problems.

Posttrip Inspection

To ensure the safety of the next driver of the vehicle, each driver should conduct a posttrip inspection. This inspection covers many of the same items as the pretrip inspection; however, it may be more extensive. When deficiencies are detected, the mechanic and the next driver of the vehicle should sign off to indicate that they are satisfied with the repairs. These records must be kept in the vehicle with all inspection records.

THE BEHAVIOR OBSERVATION PROGRAM

Objective

The goal of a behavior observation program is to create a safe work environment for employees. This program should be designed to enhance every employee's ability to recognize hazardous situations that are routinely encountered in the workforce.

Guidelines

Each employee will be asked to complete observation forms on a daily basis. The observations completed by the employees should include:

- Action of the employee
- Date of the observation
- A possible solution

The form should not include any names or identifying information. All observations completed by management should include:

- Action of the employee
- Date of the observation
- A possible solution

The form should not state any names or identifying information. In the case of hazardous actions, management will intervene immediately to prevent accidents. The observations should relate to positive safety performance as well as poor safety performance.

Information

All information should be equally shared every month with all parties involved to analyze and determine the cause of the incidents. The information should be compared to the number of service trips that day, the type of work to be completed at that time, weather conditions, and road conditions, and take into account whether the work was different from routine duties.

Discipline

The information obtained from the behavior observation program will not be applied or referenced in any pending or future discipline case. The form, when used correctly, can help in the elimination of poor driving practices that tend to cause accidents. All driving practices that are in need of improvement should be discussed with the driver as soon as possible after the observation is made. Because the primary purpose of conducting these observations is to improve driving practices before they result in an accident, discussions with drivers must be positive and include the benefits to be gained by improving driving practices. To enforce a high standard of professional driving performance, the company should observe all drivers at least twice a year, and at other times when it is appropriate. For probationary drivers, the behavior observation form should be completed a number of times during the probationary period. The manager in charge of the driver can determine the number of times it should be completed.

DETECTING HAZARDS

Interpretation

The visual search pattern is part of the Smith System, a program developed to assist drivers in identifying road scenarios. When a driver is using the visual search pattern technique, the eyes will absorb a tremendous amount and variety of information. Some of the information is absolutely essential to the driver, but most of the information will be unfamiliar. A driver must be able to interpret the information he or she sees in order to decide what to do.

Hazard Identification

The driver recognizes potential hazards in sufficient time by a combination of common sense, practice, experience, and judgment, and by using the visual search pattern constantly.

Information Processing

The process of thinking about the hazard information and planning ahead is known as *anticipation*. The driver endeavors to expect, foresee, or regard as probable what is likely to happen next.

For example, when approaching a set of green traffic signals, the driver should expect them to change to amber. This is called a "stale green light." The driver should then be prepared to stop or continue as appropriate.

Unsafe Acts

One unsafe act is the failure to secure or warn:

- Failing to place warning signs, signals, tags, and so forth
- Starting or stopping vehicles or equipment without giving adequate warning
- Releasing or moving loads without giving adequate warning
- Failing to lock, block, or secure vehicles, switches, valves, or other equipment

Improper Use of Equipment

The improper use of equipment involves:

- Using material or equipment in a manner for which it is not intended
- Overloading

EMERGENCY STOPS

Drivers should be trained to make emergency stops:

1. Press the foot brake firmly in order to stop quickly, but not so hard that the wheels lock and cause the vehicle to slide (skid).
2. Tighten your grip on the steering wheel and keep hold until the truck has stopped.
3. Just before the truck stops, press the clutch down and hold it down.
4. When the vehicle has stopped, apply the hand brake, move the gear lever into neutral, and assess the situation.

A good driver should know what is happening behind him or her at all times by checking the mirrors every few seconds while driving.

REVERSING (TERMINALS OR LOADING DOCKS)

TRUCK SPEED

Drivers should follow these directions when driving in reverse:

- Be prepared to vary the pressure on the gas pedal gently and smoothly.
- Use very little gas—about 1000 to 1500 rpms should be enough if you are on level ground.
- Use a little more if you are heading uphill or on a rough surface where up to 2000 rpms will be needed.
- Move the clutch up very slightly if you wish to increase speed.
- Move the clutch down a little if you wish to reduce speed.
- Use the foot brake to control your downhill speed very precisely.

STEERING

When reversing and turning the wheel, a driver has to allow for the time required for the steering to take effect. Because of this, the driver should begin to turn or straighten up the steering wheel slightly before it seems necessary. To enable the driver to see the effect of steering early, he or she should turn the steering wheel quickly and positively. By doing this, the driver will be able to identify movements of the steering wheel that will have the greatest effect.

OBSERVATIONS

Drivers should make the following observations:

- Check for other traffic before driving in reverse.
- Check for pedestrians and children who might be playing at the rear of the vehicle.
- Check all around before and during reversing.
- Try to make eye contact with other road users.

Drivers should always be ready to give way to other road users.

PROCEDURES FOR NEW HIRES

The Cheshire Petroleum & Gas Company implements the following procedures for new employees:

1. *Prescreening drivers*: Review the driving records of new hires to determine their competency to operate a vehicle.
2. *Driver orientation*: Drivers will be given the company's policy for vehicle usage.
3. *Vehicle familiarization*: The new employee will take a company-guided driver's test in the vehicle type he or she will be expected to operate. This allows the employee to become familiar with the vehicle and allows the company to determine the employee's competency.

PROCEDURES FOR CURRENT EMPLOYEES

The company will implement the following procedures for current employees:

1. *Continued evaluation*: Drivers must be evaluated yearly to make sure they are remaining competent. This evaluation should also be done any time an employee has an incident in which he or she is at fault.
2. *New vehicle familiarity*: Employees will be evaluated any time they are expected to operate a vehicle with which they are unfamiliar.

SUSPENSION AND REVOCATION OF DRIVING PRIVILEGES

Physical Condition

If an employee is found to be physically unable to perform the driving task safely, then the driver will have his or her privileges taken away until a qualified medical official releases that individual to return to work. An examination will be performed anytime a physical change has visibly taken place in an employee.

An employee must be able to meet the minimum standard for holding a driver's license in the state that has issued him or her a license. If the employee fails these requirements, then the company will look into revocation of that individual's driving privileges.

Unsafe Driving

An employee may have his or her driving privileges revoked if there is a noticeable history of unsafe driving. An example of this is if the employee were to have more than one incident that was his or her fault in a year. Also included are traffic violations and failure to meet physical standards.

The company is also responsible for upholding a suspension of an operator's license by the state that issued the license. If the state revokes the driver's license, then that individual will be automatically suspended from operating a vehicle for the company.

Accidents

The company will follow these procedures when an accident occurs:

1. *Review of driving privileges*: The employee's driving privileges will be reviewed by his or her supervisor anytime the employee is involved in an accident. The employee may or may not have his or her position temporarily revoked during the investigation (at the discretion of the supervisor).
2. *Assessment of circumstances*: The circumstances surrounding an accident will be assessed to include, but not limited to, the employee's condition; the seriousness of unsafe driving practices, if any, that resulted in the accident; and a determination by the supervisor as to whether the public's or the employee's safety would be jeopardized by allowing the employee to continue driving.
3. *Temporary suspension*: If an immediate determination cannot be made based on a review of the previous steps, the employee's driving privileges may be temporarily revoked pending the completion of the accident investigation. At the end of the investigation, a decision can be made to permanently revoke or reinstate the driver's privileges. This investigation has no minimum timetable but must be completed within 14 days. The maximum time of suspension without permanent revocation is 60 days.

This model of a motor fleet safety program can be used as a basis for a company's program. Company-specific information should be added as applicable.

STUDY QUESTIONS

1. (True or False) The U.S. Coast Guard was one of two DOT departments that transferred to the Department of Homeland Security in 2003.
2. (True or False) Daily in the United States, publicly funded transportation systems provide approximately 32 million passenger trips.
3. (True or False) All training should reflect security and emergency plans and procedures.
4. (True or False) One objective of crash investigations is to provide and preserve vital information.
5. According to company policy, a breath alcohol level above 0.02 percent requires the driver to be prohibited from driving any company trucks for a period of how many hours?
6. (True or False) The classroom phase of the instruction could include lectures by experts in the profession.
7. (True or False) It is important that all members of the transit workforce understand security issues and be prepared to respond should an emergency occur.
8. (True or False) Terrorist acts are not accidents or disasters, but they are intentional actions designed to inflict casualties.

16 Motor Fleet Safety and Security Management

SECURITY OPERATIONS

As a result of the events of September 11, 2001, security services have expanded operations in all aspects of motor fleet safety and security.[1] For many years, the trucking industry has worked to improve the security of trucking operations. In the aftermath of the terrorist activities, the transportation industry is expanding its security to include prevention of the use of its own vehicles and cargo (containers of explosive or combustible materials) as weapons in a terrorist attack. In June 2001, a gasoline tanker crashed, turning into a dangerous high-temperature inferno that destroyed a bridge on Interstate 80 in northern New Jersey. In another example, in a 1997 incident, an overpass was melted on the New York State Thruway.[1]

A Boeing 757 jet holds nearly 11,500 gallons of fuel, while a gasoline tank trailer can carry some 9,000 gallons.[2] Railroads carry more than 6 million gallons per year, and trucks more than 10 million. On an average day, more than 17,000 containers enter U.S. ports, and only 2 of 100 containers are actually searched or checked by proper personnel. According to the U.S. Customs Service, more than 11.2 million trucks entered the United States in 2001.[3] More than 7.5 million vehicles and approximately 10.5 million holders of commercial driver's licenses (CDLs) are widely dispersed across the country. Almost 2.5 million of those drivers have an endorsement that allows them to transport hazard materials.[4]

Defending against terrorist attacks is not new to the industry; however, recent changes in attitudes and personal values have created many new problems for motor fleet safety directors. Agencies have learned what works and, just as important, what does not work in their operating systems. Companies must implement continuous improvement and review their security procedures to determine who has access to facilities and storage areas, as well as the adequacy of protection. Carriers should know their business partners, vendors, service providers, and shippers. Security messages and training should be provided regularly to all employees. Training should be comprehensive, covering overall company security, specific procedures, and the employee's personal role in security.

In motor fleet security functions, three main areas are of utmost importance. The first is truck and cargo hijacking, which falls into the classification of theft. Vehicle hijacking is one of the most feared crimes in the United States, and members of the public who perceive themselves to be targets have high levels of insecurity. One reason for this is the apparent randomness, unexpectedness, unpredictability, and levels of violence associated with the actual crime of vehicle hijacking. The term "hijacking" specifically refers to the illegal seizure and control of a vehicle (car or truck), whereas "vehicle hijacking" means not only to seize control but in fact to rob the owner and driver of the vehicle. Traditionally, hijacking has been associated with the activities of terrorists, who often seize control of the vehicle and then use the cargo stored on the vehicle, which is often highly combustible, as a weapon.

The second main area consists of bomb threats, which cause other problems through espionage and sabotage. Bomb threats are usually received on the telephone and are generally made by individuals who want to create an atmosphere of anxiety, panic, or destruction. All bomb threats should be assumed to pose a legitimate threat to the enterprise and should be considered serious for three reasons:

- The threat of danger can cause individuals to panic; this condition can increase the risk of injury and property damage.
- The evacuation or closure of trucking facilities negatively affects employers and employees, and places a financial and operational strain on the organization.
- The fearful atmosphere and stress associated with these events can affect productivity and confidence.

The third area involves the actions trucking companies are considering for use to respond to the highest level of terrorism threat. Organizations should accumulate and organize a base of knowledge regarding the exposures of the trucking industry to terrorist acts. Also, they should learn the types of behaviors and events that can precede such an act and teach their employees to be alert and observant. When High (Condition Orange) and Severe (Condition Red) terrorism threats are declared, trucking companies must consider their own actions in parallel with those implemented by governmental agencies. These build on the procedures under the Elevated (Condition Yellow) threat that has been the baseline since 2003.

EMERGENCY ACTION PLAN

Emergencies in the motor fleet safety program could mean chaos—and chaos is hazardous. An emergency action plan describes the actions employees should take to ensure their safety if a fire or other emergency situation were to occur. People are the most important part of an emergency action plan. The implementation of an emergency plan and proper employee training (so that employees understand their roles and responsibilities within the plan) will result in fewer and less severe employee injuries and less structural damage to facilities during such emergencies. This could be a trucking terminal or a bus garage. A poorly prepared plan can lead to a disorganized evacuation or emergency response, resulting in confusion, injury, and property damage.

The best emergency action plans include employees in the planning process, specify what employees should do during an emergency, and ensure that employees receive proper training for emergencies. When including employees in planning, encourage them to offer suggestions about potential hazards, worst-case scenarios, and proper emergency responses. After developing the plan, review it with employees to make sure everyone knows what to do before, during, and after the emergency. Keep a copy of the emergency action plan in a convenient location where employees can get to it, or provide a copy to all employees.

Evacuation policies, procedures, and escape route assignments inform employees who is authorized to order an evacuation, under what conditions an evacuation would be necessary, how to evacuate, and what routes to take. Evacuation procedures often describe actions employees should take before and while evacuating, such as shutting windows, turning off equipment, and closing doors behind them. Exit diagrams are typically used to identify the escape routes to be followed by employees from each specific facility location.

Evacuation Review

All personnel should evacuate as quickly and calmly as they can and:

- Be familiar with the floor plan and layout in the department or terminal where they work.
- Know the primary and secondary exits (routes) from each location.
- Recognize that the assembly location might change relative to the emergency.
- Know where all personnel meet after evacuating the facility.
- Practice emergency evacuation drills and procedures with other departments.

Once the evacuation alarm has been sounded, the following procedures should be followed:

- All personnel should report to the designated assembly location.
- Instructions from team leaders on check-in procedures must be carried out.
- Special information must be reported at this location to all personnel.
- All personnel should remain at the assembly location until released.

HIJACKING AND CARGO THEFT

During the 1990s, hijacking of vehicles became a crime of prominence. It is a crime that in recent years has not only increased dramatically but has also been associated with higher levels of violence in its perpetration.[3] Furthermore, there has been widespread media coverage and interest in it. "The growth in these types of crimes [cargo theft and hijacking] has significantly contributed to increased levels of fear of crime [and] ... inhibits freedom of movement and economic activity and is highly costly in terms of loss of property and psycho-social damage caused by trauma and fear."[5]

Training preparedness is an ongoing concern when a company deals with vehicles and explosives, and therefore the company must address these issues on a periodic basis. In some countries, terrorists frequently use vehicles carrying explosives to destroy a target. However, terrorists have realized that tankers full of explosive fuel are convenient alternate weapons or surrogate bombs. These tanker trucks can be used against specific targets, even turning hazardous industrial facilities into chemical weapons. On June 23, 1996, in Saudi Arabia, the explosion of a fuel truck outside a U.S. Air Force installation killed 19 U.S. military personnel and wounded 515 people, including 240 Americans. In Europe, terrorists are suspected of hijacking fully loaded tanker trucks and crashing them into targets and/or detonating them with explosives when parked near a target.[2] On April 11, 2002, a loaded gas tanker truck collided with the wall of a synagogue in Tunis. The truck exploded, killing 15 people, several of them German. An Islamic terrorist group claimed responsibility. In May 2002, a fully loaded tanker truck pulled into Israel's largest fuel depot and suddenly caught fire due to an explosive charge connected to a cellular phone. The fire was extinguished, but had the truck exploded, significant destruction and death would have resulted. This shift in terrorist behavior has enhanced awareness by security specialists and has resulted in an increased risk potential for the trucking industry.

While in transit, trucks can be an easy target for violence, and companies must train drivers to be aware of their personal security and, more important, to be prepared for the unexpected. Trucks are most vulnerable when they are stopped. Few trucks in the United States are equipped with truck security systems. Technology presently exists for ignition kill capabilities and for global positioning system (GPS) tracking by fleet operators. These methods and developing technologies can play important roles over time in truck security. The trucking industry needs to develop standards to support the transition to these approaches as they become routine and available. Spot, routine, and high-risk truck inspections are infrequent. At truck stops, unattended vehicles are frequently left with their engines running, and these areas are usually unguarded. These stops require the presence of security personnel and increased law enforcement surveillance at and around these areas. There are only a minimal number of secure parking areas available along America's roadways, so drivers often use vacant unsecured areas as rest stops.

Since September 11, 2001, the role of the truck driver has become more important. Now more than ever, America depends on truck drivers to move freight safely and securely, especially if transporting hazardous materials.[6] Although high valued commodities have always been hijacking targets, shipments capable of mass destruction or environmental damage must now be protected with the utmost care.

Here are some safety tips and steps drivers should follow while traveling on the road alone with cargo, explosive, or hazardous materials. You may want to pass these along to your drivers:[2]

- Be alert when leaving. Criminal surveillance often begins at or within a mile of origin.

- Do not discuss cargo, destination, or other trip specifics on open radio channels or near unfamiliar persons.
- If you believe you are being followed, call 911 and your dispatcher immediately.
- Avoid being boxed in by other traveling or parked vehicles. Where it is possible, leave room in front and in back of your vehicle.
- Look for vehicles following you, especially if there are three or more people in a vehicle.
- If you believe you are being hijacked, try to keep your truck moving while keeping an eye out for law enforcement officers.

To stop the vehicle with a load of cargo:

- Leave the truck in a secure parking lot or truck stop if possible; if that is not possible, be certain someone can watch the vehicle in your absence.
- If team driving is involved, always leave one person behind with the truck to secure the vehicle from any suspicious activities and an attempted hijack.
- Never leave the vehicle running with the keys in the ignition; shut off the engine every time you stop and lock the doors for security purposes and safety.
- If at all possible, do not stop in unsafe or high-crime areas.
- Always lock the cargo door(s) with padlocks.
- Use seals to prevent and identify tampering.

At a minimum, motor fleet carriers need to explore and utilize low-technology security options such as the following:

- Instruct drivers to use seals, padlocks, kingpin locks, gladhand locks, ignition locks, and similar items.
- Develop and implement security policies and standard procedures (internal and external security).
- Develop security-related training and ensure that all employees are instructed in the policies and procedures of the company.
- Assess hazardous materials security risk.
- Prioritize the shipper/carrier hazardous materials operations for security.

TYPES OF EMERGENCIES AND DISASTER

Cooperation within the motor fleet security industry takes the form of individual and organized efforts when it comes to emergencies. Security professionals have the opportunity to participate in disaster education programs ranging in scope from local to national conferences. Planning and training are two areas to help reduce losses when a disaster strikes. All employees, from the drivers to the maintenance employees, need to know what to do in order to protect lives, property, and assets of the enterprise. Some of these disasters are sabotage, multiple vehicle accidents, explosions, windstorms, and floods; other natural occurrences may strike drivers on the road or in a building collapse at the company headquarters. It is essential that management and security personnel be familiar with these extremely violent and destructive forces, often occurring without any warning.

This emergency planning will increase the chances that people and an organization will survive from a disaster, whether natural or man-made. There is no one disaster plan that can be applied to all motor fleet operations that can instruct them on how to prepare and recover from a disaster. Different needs and strategies must be met. The primary point to disaster recovery aims to place an enterprise in the position it was in prior to the disaster.

BOMBS AND BOMB THREATS

Contingency plans should specify who will be responsible for handling the crises and delegating authority in the event that a bomb threat is received. Bombs can be constructed to look like almost anything and can be placed or delivered in any number of ways. The probability of finding a bomb that looks like the stereotypical bomb is almost nonexistent. The only common denominator that exists among bombs is that they are designed or intended to explode. Most bombs are homemade and are limited in their design only by the imagination of, and resources available to, the bomber. When searching for a bomb, suspect anything that looks unusual.

Bomb threats are delivered in a variety of ways. The majority of threats are called in to the target. Sometimes these calls are through a third party. Occasionally a threat is communicated in writing or on a recording. Two logical explanations for making a bomb threat are:

- The caller has definite knowledge or believes that an explosive has been or will be placed and he or she wants to maximize personal injury or property damage. The caller may be the person who placed the device or someone who has become aware of such information.
- The caller wants to create an atmosphere of anxiety and panic that will, in turn, result in a disruption of the normal activities at the facility where the device is purportedly placed.

Whatever the reason for the bomb threat, there will certainly be a reaction to it. Through proper planning, the wide variety of potentially uncontrollable reactions can be greatly reduced. The individual in charge of responding to a bomb threat must be someone who will be available 24 hours a day. In addition, all personnel who will be involved in the bomb threat response must be trained in how to handle these situations.

PROCEDURES FOR BOMB THREATS

Terrorism is a covert and criminal act that creates problems for management and emergency service personnel. Many of these acts of terrorism deal with bomb incidents, bomb threats, and the taking of hostages. In order to be prepared, companies should make contact with and hold meetings with the local law enforcement agencies, the Federal Bureau of Investigation (FBI), and bomb-disposal units. This will give a company the opportunity to gain assistance from more experienced personnel. Experience has shown that 95 percent of all bomb threats are hoaxes. This leads to the problem that a bomb threat may be ignored although it is authentic. When considering measures to increase security for buildings or vehicles, you must contact the local police department for instruction. There is no single security plan that is adaptable to all situations. The following recommendations are offered because they may contribute to reducing the vulnerability to bomb attacks. All plans should be in writing and revised when necessary.

Bombs being delivered by truck carriers or left in the carrier's cargo are a grave reality. Parking should be restricted, if possible, to 300 feet from the building or from any building in a complex. If restricted parking is not feasible, properly identified employee vehicles should be parked close to your facility and visitor vehicles parked at a distance.

Controls should be established for positively identifying personnel who have authorized access to critical areas and for denying access to unauthorized personnel. These controls should extend to the inspection of all packages and materials being taken into critical areas.

Security and maintenance personnel should be alert for people who act in a suspicious manner, as well as objects, items, or parcels that look out of place or suspicious. Surveillance should be established to include potential hiding places (cockpit area, rest rooms, and any vacancy in freight boxes or trailers) for unwanted individuals.

Responding to Bomb Threats

If at all possible, more than one person should be assigned to listen in on a bomb-threat call. To do this, a company needs to implement a covert signaling system, perhaps by using a coded buzzer signal to a second reception point. A calm response to the bomb threat caller could result in obtaining additional information. The bomb threat caller is the best source of information about the bomb. When a bomb threat is called in:

- Keep the caller on the line as long as possible. Ask him or her to repeat the message. Record every word spoken by the person.
- If the caller does not indicate the location of the bomb or the time of possible detonation, ask him or her for this information.
- Pay particular attention to background noises, such as motors running, music playing, and any other noise that may give a clue as to the location of the caller.
- Listen closely to the voice (male/female), voice quality (calm, excited), accents, and speech impediments. Immediately after the caller hangs up, report the threat to the person designated by management to receive such information.
- Report the information immediately to the police department; fire department; Alcohol, Tobacco, and Firearms (ATF); FBI; and other appropriate agencies. The sequence of notification should be established in the bomb incident plan.
- Remain available; law enforcement personnel will want to interview you.

When a written threat is received, save all materials, including any envelope or container. Once the message is recognized as a bomb threat, further unnecessary handling should be avoided. Every possible effort must be made to retain evidence such as fingerprints, handwriting or typewriting, paper, and postal marks. These will prove essential in tracing the threat and identifying the writer.

TERRORISM THREATS

As mentioned earlier, when High (Condition Orange) and Severe (Condition Red) terrorism threats are declared, trucking companies must consider their own actions in parallel with those implemented by government agencies. These build on the procedures under the Elevated (Condition Yellow) threat that has been the baseline for 2003.

Organizations dealing with motor carriers should monitor American Trucking Association (ATA) and state trucking association communications and websites closely, as well as provide information to the industry as soon as it becomes available.[7] Because events may unfold quickly and at unpredictable times, the following are some considerations in the event that High or Severe threat conditions are declared.

High Alert (Threat Condition Orange)

Follow these steps during a Condition Orange:

- Ask the management crisis team to verify team members' 24/7 contact information and place them on "ready alert" through the period of High alert.
- Reduce the internal tolerance for security anomalies, such as overdue or missing vehicles, perimeter of physical plant intrusions, unverified visitors, evidence of tampering, and the like. Report suspicious activities, especially those fitting any profiles presented in threat alert advisories, immediately to law enforcement and/or the nearest FBI field office.
- Conduct emergency/contingency procedures reviews with drivers, dispatchers, and line management personnel. Brief personnel on the threats that triggered the alert and how these threats may present themselves in the field.
- Test the emergency communications systems.
- Identify state and/or local emergency planning agencies, industry resources, and the like through which event response and recovery information can be obtained.
- Ensure that company personnel monitor news and other information sources for events or changes in conditions and respond as appropriate.

Severe Alert (Threat Condition Red)

The current interpretation of the way Condition Red may be applied is that it will be imposed based on actual information of pending attacks or in the event of an actual attack. The extreme restrictions implied under the code will likely be imposed only on specific locations, facilities, or types of operations. Actual events may determine otherwise.

If your company is not located in the areas covered by the alert or services clients located in the area covered by the alert, maintain the alert status as under Condition Orange, but monitor your information sources constantly, briefing your management and personnel on the most current information.

If your company is in the alert area or servicing areas covered by the alert, consider the following:

- Call the crisis management team to duty.
- As appropriate to the event or the threats that triggered the severe alert, alter operations according to the company plans and any specific instructions and mandates issued by governmental agencies with the authority to do so.
- Consider the advisability of "locking down" the facility to only those essential to ongoing operations and business transactions.
- If there is evidence that the company, trucks, carriers, or operations are under a direct threat or exposure to attack, make efforts to move those at risk away from the immediate threat and notify law enforcement agencies immediately.
- If the company transports critical cargo or can provide technical or logistical assistance that may be needed to respond to an event, identify yourself to local emergency response coordinators.

Security Watchwords for Corporations

Here are some security watchwords for companies:

- *Awareness*: Accumulate and organize a base of knowledge regarding the exposures of the trucking industry to terrorist acts and the types of behaviors and events that can indicate an event and teach the workforce to be alert and observant.
- *Recognition*: Train employees and managers to make the logical connections between observed indicators and a specific company's operations that may signal an imminent act or increase a company's exposure to consequences.
- *Communication*: Build a network of time-sensitive systems through which information is routed to and among the internal and external decision makers who need critical information in order to prevent or respond to terrorist actions.
- *Action*: Proactively deploy the correct measure of activity relating to the nature of the threat, the overall Homeland Security Advisory System (HSAS) threat condition level, and the trucking operation's potential exposure.

Security Watchwords for Drivers

For drivers, security watchwords include these terms defined as follows:[7]

- *Awareness*: Learn how terrorists act and the types of behaviors and events that can precede an attack. Know the company's security procedures and emergency response plans as they apply to you. Look for behaviors or events that might be a tip-off to a terrorist operation in progress.
- *Recognition*: When you see behavior changes or events that match the profiles you have been taught, make the mental connections between what you see and what it may mean to you if indeed it is a terrorist activity.
- *Communication*: Know whom to call no matter where you are. Use 911 in emergencies, and your company dispatcher and local FBI or law enforcement numbers if not an emergency.
- *Action*: Do not keep information to yourself but send it to the people and agencies who have the expertise and training to react to information or to emergencies. If you are affected by an attack, take immediate action to protect yourself, your cargo, and your equipment.

HIGHWAY WATCH PROGRAM FOR TRUCK DRIVERS

The general abilities and aptitudes required of a truck driver vary with the individual task. Every truck operator has the potential of damaging or demolishing his or her truck in an accident. Highway Watch—operated by each state's State Trucking Association (STA)—is a safety initiative that takes advantage of the skills, experience, and road smarts of America's professional truck drivers and joins the trucking industry, law enforcement agencies, and other safety professionals as allies working together to make the nation's highways much safer.[8]

The Highway Watch program trains professional truck drivers on what to look for on the highway. Training focuses on how emergencies should be reported, the appropriate telephone numbers to call, safe and responsible wireless phone use, and how the program will develop in order to coordinate with other safety and security initiatives. Highway Watch professional truck drivers receive extensive training on how to report accidents, stranded motorists, criminal and terrorist activity, medical emergencies, treacherous weather conditions, congestion, road rage, drunk driving, and potential hazards such as abandoned vehicles and road debris. Drivers will be trained to use advanced technology when they see an incident on the highway. They learn how to report to 911 in life-threatening emergencies and to use a special number for all other incidents. (The call will be transferred to the appropriate governmental agency.)

Each STA operates the Highway Watch program within its respective state, working with state law enforcement and governmental agencies to maintain and enhance the program. The STA is responsible for recruiting drivers, scheduling training sessions, and administering the program on a daily basis. As Highway Watch programs continue to expand state by state, they will also focus on establishing credibility and building relationships with national organizations whose state affiliates and leaders are central to the success of state programs.

The Transportation Security Administration (TSA) and the U.S. Department of Transportation moved to secure the transport of dangerous goods (chemicals, hazardous materials, and explosives) by issuing an interim final rule requiring background checks on commercial drivers certified to transport hazardous items. This is a landmark rule because it establishes vital safeguards to protect national transportation networks from possible acts of terrorism. The rules further ensure the continued safe transportation of a range of products, from chlorine to gasoline, crucial to the economic viability of the United States.

Under TSA's rule, the roughly 3.5 million commercial drivers with hazardous material endorsements are now required to undergo a routine background and records check that includes a review of criminal, immigration, and FBI records.[9] Any applicant with a conviction (military or civilian) for certain violent felonies over the past 7 years, or who has been found mentally or physically incompetent, will not be permitted to obtain or renew the hazardous material endorsement. The checks will also verify that the driver is a U.S. citizen and a lawful permanent resident as required by the U.S.A. Patriot Act.

The U.S.A. Patriot Act authorized the secretary of transportation to develop standards to require commercial drivers with hazardous materials endorsements to undergo a criminal history background check. The secretary delegated authority to TSA, the Federal Motor Carrier Safety Administration, and the Research and Special Programs Administration, which have worked closely with other federal agencies and industry associations to develop complementary rules addressing the transportation of hazardous materials by motor carrier fleets.

STUDY QUESTIONS

Statements 1–10 require a True or False response:

1. Vehicle hijacking is one of the most feared crimes in the United States.
2. The Highway Watch program trains professional truck drivers on what to look for on the highway.
3. In the section "Security Watchwords for Drivers," the main point is to teach drivers how terrorists act and the type of behavior events that can precede an attack.
4. When responding to bomb threats, only one person should be assigned to listen in on the calls.
5. The majority of bomb threats are called into the target by e-mail.
6. During the 1990s, hijacking of vehicles became the crime of prominence.
7. During evacuations, all personnel should be familiar with only the primary route.
8. All personnel should report to the designated assembly location when the alarm is sounded.
9. The best emergency action plans include employees in the planning process.
10. The evacuation or closure of trucking facilities negatively affects employers and employees and places a strain on the company.
11. Case Problem: If you were a company fleet safety director, how would you react to an Occupational Safety & Health Administration (OSHA) inspection of your truck terminal? What conditions in your company do you think would influence your reaction?

REFERENCES

1. Revkin, A., A nation challenged: hazardous materials—states are asked to pull over any truck allowed to carry hazardous cargo, *New York Times*, September 27, 2001.
2. Heil Trailer, *Product Specifications for Petroleum Tank Trailers*, http://www. heiltrailer. com.
3. Bigelow, B.V., Computers try to outthink terrorists, *San Diego Union-Tribune*, January 13, 2002.
4. Statement of Joseph M. Clapp, administrator, Federal Motor Carrier Safety Administration, before Senate Sub-Committee on Surface Transportation and Merchant Marine Committee on Commerce, Science, and Transportation, October 10, 2001.
5. Interdepartmental Strategy Team (Departments of Correctional Services, Defense, Intelligence, Justice, Safety and Security and Welfare), *National Crime Prevention Strategy*, 41, Government Printers, Pretoria, South Africa, May 1996.
6. Flatow, S., Safety in the trucking industry: Nondriving incidents, *Professional Safety*, American Society of Safety Engineers, Des Plaines, Illinois, November 2002.
7. American Trucking Association, *Workers' Compensation Injury Reduction and Cost Control: Strategies for the Trucking Industry*, Alexandria, Virginia, 1996.
8. National Private Truck Council, *Driver Trainer Manual*, Washington, D.C., 1981.
9. National Safety Council, *Injury Facts*, Itasca, Illinois, 2001.

Appendix A: Combustible Liquids

The classification of a material as a combustible liquid is strictly for transportation within the United States and is not recognized internationally. The term is defined in Title 49 of the Code of Federal Regulations §173.120(b)(1) as "any liquid that does not meet the definition of any other hazard class specified in this subchapter [the Hazardous Materials Regulations], and that has a flash point above 60.5°C (141°F) and below 93°C (200°F).

However, §173.120(b)(2) provides for Class 3 (flammable) materials with flash-point at or above 38°C (100°F) and up to 60.5°C (141°F) that do not meet the definition of any other hazard class to be reclassed as a "combustible liquid" for transportation by highway and rail. For shipments involving any air, water, or international movements, these materials are Class 3 (flammable) materials.

Combustible liquids in nonbulk packaging that are not a hazardous substance, hazardous waste, or marine pollutant are not subject to the hazardous materials regulations §173.150(f)(2).

Furthermore, combustible liquids in bulk packaging that are not a hazardous substance, hazardous waste, or marine pollutant are only subject to the hazmat regulations specified in §173.150(f)(3). This does not include labeling or security plans.

(f) *Combustible liquids.* (1) A flammable liquid with a flash point at or above 38°C (100°F) that does not meet the definition of any other hazard class, may be reclassed as a combustible liquid. This provision does not apply to transportation by vessel or aircraft, except where other means of transportation is impracticable.

(f)(2) The requirements in this subchapter do not apply to a material classed as a combustible liquid in a nonbulk packaging unless the combustible liquid is a hazardous substance, a hazardous waste, or a marine pollutant.

(f)(3) A combustible liquid that is in a bulk packaging or a combustible liquid that is a hazardous substance, a hazardous waste, or a marine pollutant is not subject to the requirements of this subchapter except those pertaining to

(f)(3)(i) Shipping papers, waybills, switching orders, and hazardous waste manifests;

(f)(3)(ii) Marking of packages;

(f)(3)(iii) Display of identification numbers on bulk packages;

(f)(3)(iv) For bulk packagings only, placarding requirements of subpart F of part 173 of this subchapter;

(f)(3)(v) Carriage aboard aircraft and vessels (for packaging requirements for transport by vessel, see §176.340 of this subchapter);

(f)(3)(vi) Reporting incidents as prescribed by §171.15 and §171.16 of this subchapter;

(f)(3)(vii) Packaging requirements of subpart B of this part and, in addition, nonbulk packagings must conform with requirements of §173.203s

Appendix B: Describing Hazardous Materials §172.202

Any person who offers a shipment of hazardous materials for transport must describe the hazardous materials in a specified manner.

The description of hazardous materials on shipping papers must be prescribed in the regulations. The description should be the same whether hazardous materials are alone or both hazardous and nonhazardous materials are entered.

Each hazardous material offered for transport must be clearly described on the shipping papers from the Hazardous Materials Table (HMT) of Title 49 of the Code of Federal Regulations. This shipping description must include:

1. UN/NA Identification Number (see Column 4 of the HMT).
2. Proper Shipping Name (see Column 2 of the HMT).
3. Hazard Class or Division Number (see Column 3 of the HMT).
4. Subsidiary Hazard Class(es) or Division Number(s) entered in parentheses.
5. Packing Group, if required (see Column 5 of the HMT).
6. Total Quantity, by mass, volume, or activity for Class 7 or net explosive mass for Class 1.
7. Number and Type of Packages (mandatory October 1, 2007).

BASIC DESCRIPTION §172. 202

The first five items—often referred to as material's basic description—must be shown in sequence unless authorized by the regulations. The identification number must include the letters "UN," and the packing group must be shown in Roman numerals and may be preceded by the letters "PG".

Examples:
UN1203, Gasoline, 3, PG II
UN2359, Diallylamine, 3, (6.2, 8), II

The basic description may be shown in an alternate sequence, with the proper shipping name listed first used until January 1, 2013.

Examples:
Gasoline, 3, UN1203, PG II
Diallylamine, 3, (6.1, 8), UN2359, II

The regulations allow that if a "technical name" is required, it may be entered between the proper shipping following the basic description.

Examples:

UN 1760, Corrosive Liquid, n.o.s. (caprylyl chloride), 8, PG II or UN 1760, Corrosive Liquid, n. A modifier (such as "contains" or "containing") and/or the percentage of the hazardous component may be entered.

Examples:

UN1993, Flammable liquids, n.o.s. (contains Xylene, Benzene) 3, II

TOTAL QUANTITY §172. 202(A)(5), §172. 202(C)

The total quantity of a hazardous material must be indicated. This can be by mass, volume, or activity for component Class 1. The appropriate unit of measure must be included. The total quantity is not required for hazardous material packaging containing only residue; some indication of total quantity must be shown for cylinders and bulk packagings.

Example:

10 cylinders or 1 cargo tank

The material's total quantity may be placed either before, after, or both before and after the basic description.

Examples:

1 box, 25 lbs., UN1133, Adhesives, 3, II

UN1133, Adhesives, 3, II, 25 lb. 1 box 1 box, UN1133, Adhesives, 3, II, 25lbs.

Index

V

W